Everything Coming Out of Nothing

vs.

A Finite, Open and Contingent Universe

Manuel M. Carreira S.J.

U. Pontificia de Comillas
Madrid

&

Julio A. Gonzalo

U. San Pablo CEU
Madrid

CONTENTS

CHESTERTON ON "THE WHOLE EVOLUTIONARY COSMOS"

"For those who <u>really think</u>, there is <u>always</u> something really unthinkable about the whole evolutionary cosmos, as they [<u>the Evolutionists</u>] conceive it; because it is <u>something coming out of nothing</u>; an ever-increasing flood of water pouring out of an empty jug... In a word, the world does not explain itself, and cannot do so merely by continuing to expand itself. But anyhow, it is absurd for <u>the Evolutionist</u> to complain that it is unthinkable for an admittedly <u>unthinkable God</u> to make everything out of nothing, and then pretend that it is <u>more thinkable</u> that <u>nothing</u> should turn itself into <u>everything</u>". (Emphasis added).

G. K. Chesterton (1874 - 1936)
"St. Thomas Aquinas ", pp. 215 -16
(New York: Sheed and Ward, 1933)

FOREWORD

The book by Manuel María Carreira, SJ, and Julio A. Gonzalo, is a fascinating survey of the intellectual deviation of one of our great scientists, highly publicized, Steve Hawking. The subject is particularly sensitive because of the fascination which from eternity, stars and the universe in general operates on humanity. This goes as far as the Babylonians or Chinese of the same period. We know what scientific impact had on the study of planetary orbits in particular in the work of Newton, the great predecessor of Hawking at Cambridge. To understand this deviation, we must go back to the tragedy that occurred in the 19th century with the Naturphilosophie, high priests were Hegel and Schelling. The divorce between philosophy and science had been consumed by then. As a new Descartes who asked the entire Europe to communicate scientific results available to him so that he could interpret them in the light of the subjectivity born from his cogito, Hegel created a science only submitted to his dialectic which was highly unscientific. Whether the supporters of the German Naturphilosophie have dedicated to public obloquy Science, to recall the own words of Schelling "The method of experimental physics is worthless and futile, false in its principles, and is an inevitable and eternal source of errors", it was obnoxious to separate science from philosophy.

Speaking of the scientific incompetence of Schelling and Hegel, Gauss could not help asking the astronomer Schumacher "their definitions, do not make your hair stand on end?". Helmholtz in 1862, summed up the crisis that started: "The philosophers accused the scientific men of narrowness; the scientific men retorted that the philosophers were crazy. And so it came about that men of science began to lay some stress on the banishment of all philosophical influence from their work; while some of them including men of the greatest acuteness went so far as to condemn philosophy altogether, not merely only as useless but as mischievous dreaming. Thus it must be confessed not only were the illegitimate pretensions of the Hegelian system to subordinate to itself all other studies rejected, but no regard was paid to the rightful claims of philosophy, that is, the criterion of the sources of cognition and definitions of the functions of the intellect".

The wealth of discoveries in physics led the mathematician Harnack to say in a conference in Berlin in the late 19th century "People complain that our generation does not have any more philosophers. It's true: it's just that the philosophers of our time are in another department and their names are Planck and Einstein "while Heisenberg advised the young Weisäcker who planned to do philosophy. "Nobody can go anywhere whatsoever in philosophy today without knowing something in contemporary physics".

It is not surprising that it was mathematicians who founded the philosophy which underlies contemporary thought. Whether Husserl, who defended a thesis in mathematics with one of the greatest master in this discipline, Weierstrass, or Whitehead who was with Russell the writer of the Principia Mathematica, a logical compendium of the whole mathematical science, hoping to base permanently this science on intangible bases. The imperia of mathematics, on which contemporary thought Gian-Carlo Rota highlighted the pernicious influence in philosophy, will obviously be reflected in the thinking of Hawking who is basically a mathematician who describes phenomena related to stars. Black holes as evidenced with another mathematician R.Penrose – he who created the famous pavement which is grist to the mill to Gödel's theorem - are in fact singularities of solutions of differential equations!

To have everyone agree, we can only mention Etienne Gilson (1884-1978) "Nothing beats the scientific ignorance of philosophers to that, in parallel, of scientists in the field of philosophy".

The settling of accounts vis-à-vis the Catholic Church relies on biased data on Galileo's trial, which was publicized mainly in the 19th century -not to mention the sentence falsely attributed to Galileo by Bertolt Brecht "and yet it turns!" - Nowadays, the case « Giordano Bruno » is more trendy…It is fascinating how an itinerant true pathology who never ceased to be kicked out of the various kingdoms of his time because of intellectual dementia is turned to a matyrdom because of Church obscurantism !

The position of the Church is founded on realism. G. Chesterton said that only she could retain a strong sense of reality that lacked much in view of the contemporary intellectual pollution. This is what has been claimed since Cardinal Barbieri asked Galileo if he could say that his assumptions were the only ones that could account for observed phenomena, to Blessed Pope John Paul II pointing out that science is only a model of reality and cannot be exhausted.

Thus this contemporary crisis is fundamentally materialistic and has its roots in the defection of a realist philosophy vis-à-vis the spectacular results of modern science. Even Phenomenology, which is the least toxic of the modern philosophic thinking is not interested in the being as such but in the way the being is grasped. Definitely, this is not realism. The least materialist among nowadays scientists we can meet would be for example Einstein - but only at the end of his life...- who was, for the better, "a spinozist of the becoming" to use another phrase of Etienne Gilson. No wonder that a Creator subtly emerging from contemporary astrophysics would turn crazy materialists of any kind, Hawking first. Disappearance of Carl Sagan in further editions of the best seller "A brief history of time" is also to be pointed out: before he died in December 1996, he confessed his belief in God turning away from the materialist vision he expressed in the foreword of S. Hawking's book.

In conclusion, the book by Manuel María Carreira, SJ, and Julio A. Gonzalo, brings in the field of astrophysics an important diagnostic on how these contemporary mental illnesses are no longer able to distinguish the real from its scientific modeling or even the metaphysics which they insinuate. To explain to the lay man what is being made in laboratories without making a wolf in sheep clothing is a challenge. A real makeover for the goodwill reader.

Professor Jacques Vauthier

Mathématiques, Sorbonne
Paris VI

PREFACE

Some contemporary scientists do not care about the principle of energy conservation and say seriously that the universe comes out of nothing. Of course Planck, Einstein and Lemaitre would disagree. But they do not seem to worry in the least. This is a clear proof that nowadays nonsequiturs are easily accepted in academic circles as most serious scientific statements.

After an introduction in which the authors review high level presentations at the 1993 Summer Course in El Escorial in which two future Nobel Prize winners, John C. Mather and George F. Smoot were principal speakers, this book puts into perspective the views of Stephen Hawking, from his presentation at the Vatican Study Week on Astrophysical Cosmology (1981), his book "A brief history of time", (1988), his lecture on "Gödel and the end of physics" (2002), and his latest book "The Grand Design" (2010).

Chapters on the origin of science in the Christian West, the Post-Renaissance Revolution and the true pioneers of Modern Physics follow. A concluding chapter reviews briefly the evidence for a finite, open and contingent universe, and an illuminating Appendix on "The Chaos of Scientific Cosmology", by the late historian and philosopher of science Fr. Stanley L. Jaki, recently deceased, completes the book.

In summary, pretending that "Everything coming out of nothing" is a realistic description of the origin of the universe is not logically, epistemologically or metaphysically tenable. On the other hand "A finite, open & contingent universe" is a reasonable coherent and intelligible description.

A description well grounded on the scientific evidence now available.

In pre-Christian civilizations (India, China, Egypt, Mesopotamia, even pre-Christian Greece and Rome), as noted by Stanley L. Jaki, the cosmos was viewed as an eternal, infinite entity, subject to eternal returns. Only in medieval Catholic Europe, under the protecting shadow of right reason aided by Biblical revelation, did the view of a finite, open and contingent universe emerge. To some extent, this Christian view is shared also by Jews and Muslims philosophers who did not relapse to pagan pre-Christian pantheism.

Manuel M. Carreira S.J.
U. Pontificia de Comillas
Madrid

&

Julio A. Gonzalo
U. San Pablo CEU
Madrid

CHAPTER 1

Introduction

Manuel M. Carreira S.J.[1] and Julio A. Gonzalo[2]

[1]*Universidad Pontificia de Comillas and* [2]*Escuela Politécnica Superior, Universidad San Pablo CEU, Madrid*

Abstract: A brief summary of recent developments in theoretical cosmology is given. The impact of Professor Hawking's "A Brief History of Time" is put into perspective. Lectures on the subject by Professor Jaki, at a Symposium in Madrid (1990), a Summer Course at El Escorial (1993) and at the UAM, Madrid (2002) are introduced.

Keywords: Astrophysical Cosmology, Stephen Hawking, Vatican Study Weeks, Summer Courses at El Escorial, Ralph Alpher, John C. Mather, George F. Smoot, Stanley L. Jaki.

Stephen Hawking is, undoubtedly, one of the best known and most celebrated media personalities of our time. Proof of that is that he can fill large amphitheaters at international scientific events and that people is willing to pay ten or twenty dollars for attending his talks. One of the authors of this book (JAG) was at one such event in Atlanta (the Centenary of the American Physical Society, 1998) with a younger colleague. They were thinking about attending Hawking's conference. When they saw the price they found it a little too much.

Stephen Hawking, occupant today of the prestigious Lucasian Chair at Cambridge University (previously occupied by Newton, Maxwell and Dirac among other luminaries) is a very bright mathematical physicist. Among other things, he has contributed, in collaboration with the British mathematician Roger Penrose, to the theory of black holes. His contributions, consisting of clever considerations about the entropy and temperature of those mysterious entities, are, for the time being – and probably for many years to come- far from being observationally testable. In this respect, the work of Stephen Hawking is not comparable to the work of Albert

*Address correspondence to **Manuel M. Carreira:** Universidad Pontificia de Comillas, Madrid; Tel 34-91-540-6101; Fax 34-91-372-0218; E-mail: ecarreira@res.upco.es

Einstein. Einstein's General Theory of Relativity was tested experimentally within few years of its publication in 1916, when the expedition of Sir Arthur Eddington to South Africa confirmed that light rays from a distant star are gravitationally bent when passing near the Sun's surface. Einstein became overnight a world celebrity. But then he said something to the effect that if his theory had not been confirmed experimentally it deserved to be forgotten as no more than a beautiful speculation. However, it was confirmed. Therefore, it deserved the crisp contemporary comment of Bernard Shaw: "Einstein has not challenged the facts of science but the axioms of science, and science has surrendered to the challenge". (See f.i., "Thirty years with G.B.S", London: V. Gollancz, 1951, p. 194).

In this book, we put in perspective some relevant publications of Stephen Hawking since 1981, the year in which he attended, together with a very select international group of cosmologists, the Vatican Study Week on Astrophysical Cosmology at Rome sponsored by the Pontificia Academia Scientiarum. That year, Pope John Paul II, barely recovered from the shots at him by Ali Agca, addressed warmly the participants of the Study Week and, according to Hawking, dared to say that "it was all right to study the evolution of the universe after the big bang, but (that) we should not inquire into the big bang itself, because that was the moment of Creation and therefore the work of God" ("A Brief History of Time", p. 116).

What John Paul said actually to the participants in the Study Week (directly translated from the French original) was the following: "Every scientific hypothesis about the origin of the world, like that of a primordial atom from which the ensemble of the physical universe flows, leaves open problems concerning the beginning of the universe". Science cannot solve by itself such a question: it is necessary man's wisdom, which is capable of going over from the physics and the astrophysics to that which one calls the metaphysics; it is necessary too, above all, the wisdom coming from God's revelation. Thirty years ago, on November 22, 1951, my predecessor Pope Pius XII, speaking on the origin of the universe at the time of Study Week on the Problem of the micro-earthquakes, organized by the Pontifical Academy of Sciences, said:

"(To the question of the age of the cosmos). An answer is not to be expected from the natural sciences, which should admit candidly, on the contrary, that for them it is an insoluble enigma. It is equally true that, asking (that question) from the natural sciences, as such, might be asking too much. It is undeniable too, that an intellect illustrated and enlightened by modern scientific knowledge, after considering the problem with equanimity, is driven to break the (fatal) encirclement of an autonomous and totally independent matter –either uncreated or self created– and ascend up to a creating Spirit. With the same clear and analytical perspective with which that intellect examines and judges the facts, it perceives and recognizes the work of a creating Omnipotence, whose virtue puts into motion the potent "fiat" pronounced billions of years ago by the creating Spirit, and displays itself in the universe, calling into existence matter overflowing with energy with loving and generous love".

Figure 1: El Escorial, 1993.

The poetic overtones of the above words then recalled by Pope John Paul II may have not been to the liking of Professor Hawking but, in all likelihood, they were not intended as a direct prohibition to inquire into the Big Bang itself.

As the photos from the Study Week Proceedings show, Professor Hawking, his children Timothy and Lucy and Mrs. Hawking (Jane Wilde) were treated very well, indeed, at the Vatican, during that Study Week.

In 1988, Hawking published his "A Brief History of Time". Apparently, he did not expect such a great success: It was in the London "Saturday Times" bestseller list for over four years, longer than any other book has been all these years. Soon it was translated to other languages, and it is estimated that about nine million copies have been sold worldwide. Now and then people kept asking him when he would write a sequel. He did so years later, in "The Universe in a Nutshell", a well illustrated and attractively presented book, which was intended to have a tree structure: Chapters 1 and 2 providing the central trunk, the others Chapters branching off. The Illustrations in the book provided an alternative route to the text, and the boxes were intended to provide the reader with an opportunity to delve into a number of advanced topics like 11-dimensional supergravity, P-branes, M-theory, Quantum Mechanics, General Relativity, 10-dimensional membranes, Superstrings, Black Holes and the like.

Stanley L. Jaki (1924-2009), probably one on the greatest historians of science in the twentieth century, sent one of the co-authors of this book (JAG) a copy of his comments on Hawking's book, published in "Reflections", Vol. I, No. I, Spring 1988. They are reproduced in this book later (Chapter 3).

Shortly afterwards, Gonzalo was invited by his good friend Professor Manuel Tello to the examining committee of a PhD Dissertation at the University of Bilbao, Spain. There, after the dissertation was over, at lunch time, Gonzalo was seating next to Professor Pedro Echenique. He was well aware of the success of "A Brief History of Time" and Gonzalo gave him at the time a copy of Jaki's article in "Reflections", provocatively entitled "Evicting the Creator". Echenique liked it very much and promised to help with Bank Bilbao-Vizcaya to get financial support to bring Jaki to talk to Madrid.

He did, and Gonzalo managed to organize a Symposium at the "Real Sociedad Económica Matritense" on October 26 and 27, 1990, with Professor Jaki giving the opening lecture on "Physics and the universe: From the Sumerians to the late 20th century".

In the years to come, specially since 1993, the year of a Summer Course on Astrophysical Cosmology in El Escorial [1], at which Ralph Alpher, John C. Mather and George F. Smoot, with Stanly L. Jaki (among other distinguished physicists) were present, Gonzalo had numerous occasion to interact with Professor Jaki, in Madrid, in Philadelphia and even Mexico City, only about one year before his untimely death.

Jaki's views on modern cosmologists, including Stephen Hawking, are given at length in "God and the cosmologists", first published in March 1989, and then reprinted, with an introduction and an extended Postscript, in 1998.

In the mid nineties Pedro Echenique, who was a PhD from Cambridge University, invited Gonzalo to come to San Sebastian for an academic session at which the keynote lecturer was Professor Anthony Hewish, a Physics Nobel Prize winner from Cambridge, co-discoverer of the "pulsars", who spoke about the new developments in Cosmology. At the end of the academic session, Professor Hewish, her charming wife (Professor of Music at Cambridge), a group of basque scientists, friends of Echenique, and Gonzalo, were invited to dinner. During the dinner, Professor Hewish was sitting next to Pedro, and they were surrounded by basque colleagues (speaking "euskera" all the time among them). Gonzalo was in front of Mrs. Hewish, and she asked him whether he played some musical instrument. He told her "sorry, only the harmonica (when I was a young boy)". "Very nice", she said, "so you could carry it conveniently in your pocket". Occasionally Gonzalo interchanged a few words in English with Professor Hewish and with Echenique. Hawking's book (and Jaki's criticism) came about in the conversation, and Gonzalo remembers hearing to say that Hawking was very arrogant indeed, but that Jaki's was no less arrogant.

At that time Gonzalo had just come to the conclusion that during cosmic evolution the time at which matter and radiation energy densities were equal did coincidence precisely with atom formation time, *i.e.*, the time at which matter and radiation decouple. He remembers also that Prof. Hewish caught the potential significance of that coincidence immediately. They corresponded thereafter about the matter. Gonzalo tried to see him few years later at Cambridge, when he came to visit Beatriz Noheda and Mike Glasser in Oxford, but Professor Hewish had some other commitment and it was not possible to see him at that time.

Some years later, in 2002, one year after Islamic terrorist wiped out Manhattan's Twin Towers from New York's skyline, Jaki was again in Madrid, at Gonzalo ´s invitation. He gave two talks. One, at Santa Rita Parish, near Moncloa, where he spoke on "Islam, Christianism and Science". The other, at Universidad Autonoma de Madrid, on "Gödel's incompleteness theorems". At his conference Jaki commented upon Hawking's recent lecture on the same subject at Dirac's centennial celebration in Cambridge.

And just very recently, Hawking's new book "The Great Design", 2010, was published. Spanish newspapers, specially "El Pais", took notice of it, as if the book were a final confirmation that creation without a Creator was possible.

Since 1988, Hawking's best question: "What is it that breathes fire into the equations and makes a universe for them to describe?" ("A Brief History of Time", p.174) remained unanswered. Who did? Albert Einstein ?: No. Alexander Friedmann ?: No. George Lemaitre ?: No. Hubble ?: Gamow?.Penzias and Wilson ?: No.

The fire is there. And somebody put it there. The universe definitely is real, rational and consistent.

Not only did the universe needs a Creator to come into existence out of nothing. It needs also Somebody to keep it, consistently, into existence all the way thereafter.

As Jaki notes, Hawking's books are full of philosophical questions, but the answers are invariably postponed.

In an interview at the Sunday Times Magazine, June 19, 1988, Jane Wilde (Mrs. Hawking) said:

"There is an aspect of his thought I find increasingly upsetting and difficult to live with. It is the feeling that, because everything reduces to a rational mathematical formula, that must be the truth…

You can't actually get an answer out of Stephen regarding philosophy…

A common sense observation.

CONFLICT OF INTEREST

The author(s) confirm that this chapter content has no conflicts of interest.

ACKNOWLEDGEMENT

Declared none.

REFERENCES

[1]   Gonzalo J.A., Sanchez Gómez J.L., Alario M.A. (Eds) "Cosmología Astrofísica" and Brük H.A., Goyneand G.V. Longair M.S. (Eds) "Astrophisical Cosmology" (Pontificiae Academiae Scientiarum Scripta Varia: Vatican City, 1982).

CHAPTER 2

At the Vatican Study Week on Astrophysical Cosmology, 1981

Manuel M. Carreira S.J.[1] and Julio A. Gonzalo[2]

[1]*Universidad Pontificia de Comillas and* [2]*Escuela Politécnica Superior, Universidad San Pablo CEU, Madrid Spain*

Abstract: The Vatican Study Week on "Cosmology and Fundamental Physics, October 2, 1981 to which S. Weinberg, D.W. Sciama, J.E. Gumn and S. W. Hawking and Ya. B. Zeldovich contrinuted is introduced. The introductory remarks by H.A. Brück are reproduced in full as well as a summary of Professor's Hawing communication "The boundary conditions of the universe".

Keywords: Astrophysical Cosmology, S.W. Hawing, H.A. Brück, Boundary conditions of the universe.

In September 28 –October 2, 1981, Vatican Study Week on "Cosmology and Fundamental Physics" was held at Casina Pius IV, seat of the Pontificial Academy of Sciences.

Professor Carlos Chagas, President of the Pontifical Academy of Sciences [1] welcomed the distinguished participants. In the preface of the Proceedings he gives a brief introduction about the history of the Academy and about the warm welcome of His Holiness Pope John Paul II, who, few months before had been seriously wounded at Saint Peter Square during the public papal audience, in a dramatic terrorist attempt against his life. The rumors that higher up Soviet authorities were involved in the undertaking were never convincingly demonstrated to be false.

The Academy's origins go back to the foundation in 1603 by Federico Gesi of the "Academia dei Lincei" (The Academy of the Linxes) which counted among its members Galileo Galilei.

*Address correspondence to Manuel M. Carreira S.J. and Julio A. Gonzalo: Universidad Pontificia de Comillas and Escuela Politécnica Superior, Universidad San Pablo CEU; Tel: 34-91-547-0815; UAN: 34-91-497-4767; Fax: 34-91-497-8579; E-mails: ecarreira@res.upco.es and Julio.gonzalo@uam.es

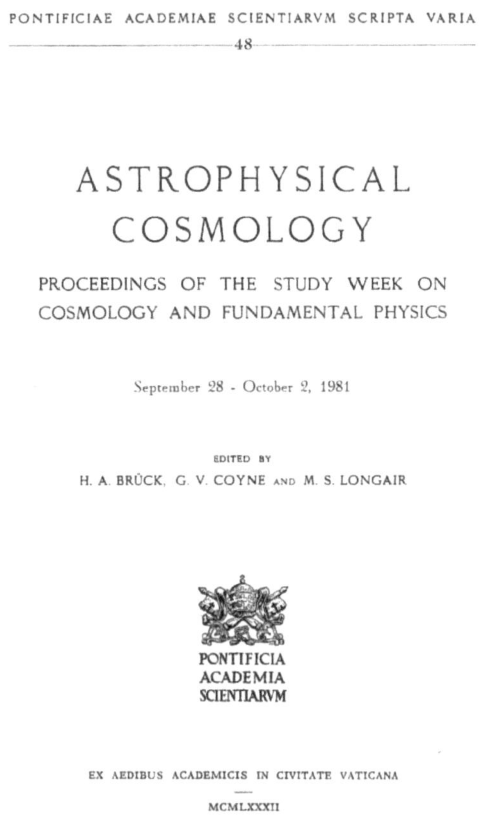

PONTIFICIAE ACADEMIAE SCIENTIARVM SCRIPTA VARIA
————————————48————————————

ASTROPHYSICAL
COSMOLOGY

PROCEEDINGS OF THE STUDY WEEK ON
COSMOLOGY AND FUNDAMENTAL PHYSICS

September 28 - October 2, 1981

EDITED BY
H. A. BRÜCK, G. V. COYNE AND M. S. LONGAIR

PONTIFICIA
ACADEMIA
SCIENTIARVM

EX AEDIBUS ACADEMICIS IN CIVITATE VATICANA
—
MCMLXXXII

Figure 1: Cover of the Proceedings of the Study Week on Cosmology and Fundamental Physics.

In 1847 the Academy was reformed by Pope Pius IX and became an official part of the Roman Sate. Pius XI, in 1936, transformed the "Academia dei Nuovi Lincei" into the Pontifical Academy of Sciences. Since then, the Academy is made up of a reduced number of distinguished international scientists, elected by the membership on the basis of scientific merit, with no discrimination regarding religion, race or nationality.

Max Planck and Albert Einstein were among the most distinguished members of the Pontifical Academy for many years. The 1981 Study Week on Astrophysical Cosmology was organized by Academicians Hermann A. Brück and Fr. G.V. Coyne, S.I., together with Prof. Martin Rees of Cambridge and Prof. Malcolm Longair of Edimburg.

Figure 2: Participants of the Study Week at the Pontifical Academy (1981).

INTRODUCTORY REMARKS BY H.A. BRÜCK

This conference has been opened by Professor Chagas, the President of the Pontifical Academy, who has spoken to you about the Academy's work and in particular about the institution of its Study Weeks.

I should like to join him first of all in welcoming this distinguished body of scientists. We are very happy that so many of you have been able to come. We are extremely sorry that none of our Soviet colleagues could accept the Academy's invitations, but we are pleased to have received a paper from Professor Zeldovich which will be read to us by Professor Martin Rees. We much regret that Professor Freeman Dyson who intended to take part was taken ill on his journey here and cannot be with us.

Perhaps I may be allowed to say a few words about the way in which this particular Study Week has come about. You have heard from Professor Chagas that Study Weeks devoted to a variety of topical scientific subjects were started by the Pontifical Academy in 1948 as a major part of its activities. Two of them were directly concerned with astronomy. In 1957 we had a Study Week on the problem of Stellar Populations, and in 1970 we had a second one in which Nuclei of Galaxies were discussed. Both of these arose from the initiative of Father O'Connell, the former Director of the Vatican Observatory, who, you will be sad to hear, is unfortunately too ill to be here with us this morning. Several of you have attended one or both of those earlier Study Weeks which, I believe it would be right to say, have been eminently successful.

It seemed to me that the time had come for a third astronomical Study Week, and the field of Cosmology appeared to present a particularly appropriate subject for this Academy. Apart from its obvious topical interest, a Study Week on present-day cosmology would be a tribute to one of the earliest members of this Academy and its President from 1960 until his death six years later. I am speaking, of course, of George Lemaitre who has been called by one of our participants in this Study Week "the Father of Big-Bang Cosmology".

When the Council of the Academy had formally agreed to hold this particular Study Week, I thought it best, not being an expert in the field, to consult with Sir Martin Ryle, an old friend and fellow member of the Academy. Martin Ryle suggested that I should discuss the project with Malcolm Longair, his close colleague who at the time was still in Cambridge. I also approached Martin Rees at the Institute of Astronomy in Cambridge, and it was he who made the important suggestion that it would be interesting and useful if the proposed subject of cosmology would be linked with that of fundamental physics. Our small group was joined by Father George Coyne, the present Director of the Vatican Observatory, who undertook to look after many of the practical problems including the recording of the scientific discussions in the course of this week. Father Coyne has also agreed to be responsible for the important task of the eventual printing of the proceedings.

Since February of this year when Coyne, Longair, Rees and myself met in Edinburgh, the major part of the preparation and planning of the actual scientific programme has been in the hands of Malcolm Longair, and a look at the programme in front of you shows you how splendidly he has performed his task.

Here in Rome the innumerable detailed preparations necessary for the smooth running of our Study Week have been most ably carried out by Father Enrico di Rovasenda, O.P., the Director of the Academy's Chancellery and Mme. Michelle Porcelli Studer and their staff. I am certain that we shall all be very well looked after during the week.

You see from the scientific programme that you will be fairly hard worked in nine sessions during four and a half days. There will be a break on Wednesday afternoon when you will have a chance to visit the Vatican Museums. On Saturday morning there will be an audience of Pope John Paul II for both the participants in the Study Week and the members of the Pontifical Academy who will later start their biannual Plenary Session. The audience will be at the Pope's summer residence at Castel Gandolfo in the Alban Hills where you will also have a chance to visit the Vatican Observatory.

We are now probably ready to start the scientific programme and I ask Malcolm Longair to take the Chair for the first Session.

*   *   *

In the next to the last session of the Study Week (VI. The Very Early Universe and Particle Physics), S. Weinberg, D.W. Sciama, J.E. Gunn and S.W. Hawking presented their respective communications. Prof. Ya. B. Zeldovich, from the Space Research Institute, USSR Academy of Sciences, Moscow, was unable to come to Rome but his communication on "Spontaneous Birth of the Closed Universe and the Anthropic Principle" was included in the Proceedings.

A brief summary, in his own words, of the communication by Prof. Hawking, entitled "The boundary conditions of the universe", follows:

> *"This paper considers the questions of what are the boundary conditions of the universe and where should they be imposed. It is difficult to define boundary conditions at the initial singularity and, even if one could, they would be insufficient to determine the evolution of the universe. In order to overcome this problem it is suggested that one should adopt the Euclidean approach and evaluate the path integral for quantum gravity*

over positive definite metrics. If one took this metrics to be compact, one would avoid the need to specify any boundary conditions for the universe. This approach might explain why the apparent cosmological constant is zero, why the universe is spatially flat, and why it was in thermal equilibration at early times".

The considerable mathematical skills of Prof. Hawking were put to work there and then to advocate a Euclidean approach to quantum gravity in order to get rid of the need to specify boundary conditions for the universe.

Thirty years later, doubts and inconsistencies are still surrounding the zero or non-zero value of the cosmological constant, the flatness or openness of the universe and its remarkable degree of thermal equilibrium at the beginning of time. All this, in spite of ever renewed theoretical speculations, including new versions of the inflationary theory and the multiverse theory.

CONFLICT OF INTEREST

The author(s) confirm that this chapter content has no conflicts of interest.

ACKNOWLEDGEMENT

Declared none.

REFERENCES

[1]   Brück H.A., Coyne. G.V. and Longair M.S. (Eds) "Astrophicial Cosmology" Pontificiae Academiae Scripta Varia: Vatican City, 1982.

CHAPTER 3

Hawking on "A Brief History of Time", 1988

Julio A. Gonzalo[*]

Escuela Politécnica Superior, Universidad San Pablo CEU, Madrid

> **Abstract:** The successive editions (1988, 1996, 1998, 2005) of "A Brief History of Time" are briefly commented upon. Some big questions such as "Where did me come from?", "Why is the universe what it is?", brought forward but left unanswered by Professor Hawing, are pointed out. A "complete unified theory of physics" is within the reach today's theoreticians, according to him.

Keywords: Wormholes, unified theories of physics, rational laws governing the universe, cosmic background radiation, COBE, CBR anisotropies, self-contained universe with no beginning and no end.

"A Brief History of Time" [1], subtitled "From the Big Bang to Black Holes", written by Stephen Hawking and first published by Bantan Dell Publishing Group in 1988 has been unquestionably the most successful popular science book of all history. Since that year it has gone through numerous editions:

- 1988 Edition. Included an Introduction by Carl Sagan, for some reason not printed in successive editions.

- 1996 Edition. An illustrated, updated and expanded edition. It contained full color illustrations and photographs and some additional topic not covered in the 1988 edition.

- 1998 Edition (Tenth Anniversary Edition). The same as the 1996 edition. In paperback and with only a few diagrams included.

- 2005 (Special) Edition. "A Briefer History of Time" (an abridged version made in collaboration with Leonard Mlodinov).

*Address correspondence to Julio A. Gonzalo: Escuela Politécnica Superior, Universidad San Pablo CEU, Madrid; Tel: 34-91-547-0815; UAN: 34-91-497-4767; Fax: 34-91-497-8579; E-mail: julio.gonzalo@uam.es

According to Hawking, the success of his book indicates the interest of the general public for big questions such as:

Where did we come from?

Why is the universe what it is?

In the Tenth Anniversary Edition Hawking took the opportunity to update the book. He included, for instance, a new chapter on "wormholes" – little tubes that connect different regions of space-time. If they can be created and kept in place - Hawking assures us- "wormholes" can be used to travel at high speed through the galaxy, or to travel back in time to another epoch, f.i. to the times of the Roman Empire, or into the future, say to the XXV century. He discussed possible explanations for this.

He also describes scientific progress in "dualities" or "correspondences" between apparently different theories of physics. According to him, these correspondences are strong indications that there is a complete unified theory of physics. He notes however that it may not be possible to express this theory in a single fundamental formulation. He thinks that we may have to use different reflections of the underlying theory in different situations. That would produce a revolution in our view of the unification of the laws of the physical world, but this would not change, according to him, the most important point: "that the universe is governed by a set of rational laws that we can discover and understand". In this statement (at least in this statement) Hawking seems to be in agreement with the greatest minds of all times: (Plato, Aristotles, Saint Augustine and Saint Thomas Aquinas): the universe is intelligible. It is well done. And human intellect is well suited to investigate it, in spite of its obvious shortcomings.

It is easy to conclude from these two facts (the universe's intelligibility and man's genuine intellectual capability to investigate it) that both are made one for the other, that they are created. And therefore that there is a Creator.

According to Professor Hawking, writing in the Tenth Anniversary Edition of his "A Brief History of Time", on the observational side (by far the most important) since the publication of the 1st edition, we had the result of COBE, the Cosmic

Background Explorer, with their beautiful confirmation of the Planck character of the CBR, and the long time expected minute CBR anisotropies. As Hawking himself says, perhaps echoing George Smoot, these fluctuations are "the fingerprints" of the creation.

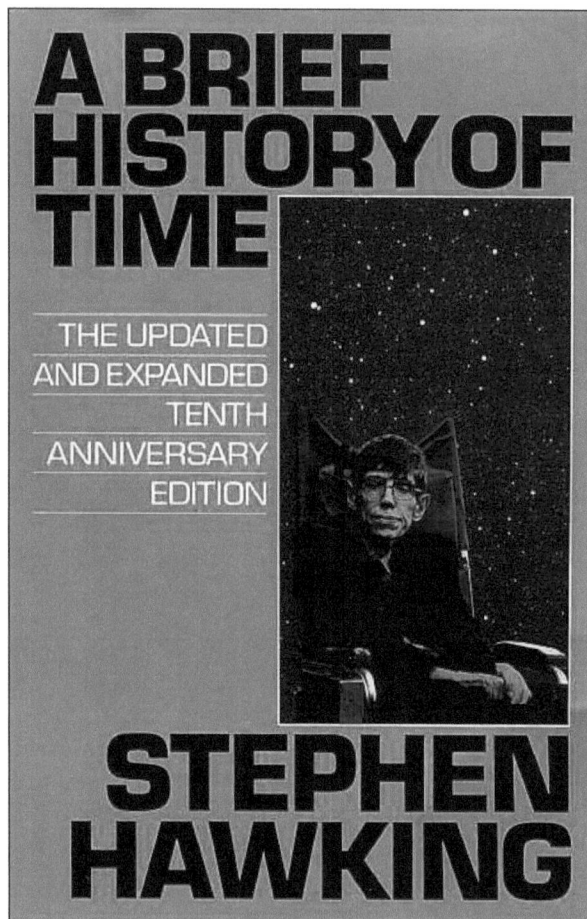

Figure 1: Cover of The Tenth Anniversary Edition of "A Brief History of Time".

He ends up, however, saying that within a few years (it is always within a few years) we should know whether we can believe that we live in a universe that is "completely self-contained and without beginning or end".

CONFLICT OF INTEREST

The author(s) confirm that this chapter content has no conflicts of interest.

ACKNOWLEDGEMENT

Declared none.

REFERENCES

[1]   Hawking. S. "A Brief History of Time" Bantam Books: New York, 1988.

CHAPTER 4

Hawking on "Gödel and the End of Physics"

Julio A. Gonzalo[*]

Escuela Politécnica Superior, Universidad San Pablo CEU, Madrid

Abstract: A brief summary of Professor Hawing is lecture is given at the Center of Mathematical Science, Cambridge University, July 20, 2002, entitled "Gödel and the end of physics". An overview of the triumphs of mathematical physics from Newton to t'Hoff is followed by the final statement that it may not be possible to formulate a theory of the universe in a finite number of statements, which is reminiscent of Gödel's theorem.

Keywords: Newton's "Principia Mathematica", scientific determinism, A.H. Compton on man's freedom, finiteness of the universe, Gödel's theorems.

At the centenary celebration of Dirac's birthday [1], in the Centre of Mathematical Sciences at Cambridge University, on July 20, 2002, Professor Stephen Hawking lectured on "Gödel and the end of physics".

The lecture is, of course, the intellectual property of Professor S. W. Hawking, and may not be reproduced, edited or translated in any way without the permission of Professor Hawking. One hopes, however this does not mean that, in order to make a critical comment, a very brief summary of its contents is not allowed.

Before discussing the relevance of Gödel's theorems for physics, Professor Hawking paints a beautiful landscape with the triumphs of mathematical physics from Newton's Principia Matematica to the end of the twentieth century. Laplace, Maxwell, Dirac, Schrodinger, Heisenberg, Abdus Salam, Weinberg, Gross, t'Hoff... appear in quick succession on stage, as actors contributing important insights to nature's understanding and nature's knowledge.

How far can we go in that search? Will we ever find a complete form of the laws of Nature? Professor Hawking asks himself.

*Address correspondence to Julio A. Gonzalo: Escuela Politécnica Superior, Universidad San Pablo CEU, Madrid; Tel: 34-91-547-0815; UAN: 34-91-497-4767; Fax: 34-91-497-8579; E-mail: julio.gonzalo@uam.es

He says, for instance, that Newton is his "Principia Mathematica" (1687) was the first one to make the laws of physics quantitatively precise. This was not exactly what happened: Almost one hundred years before, Galileo has enunciated the law connecting space's length and time duration in the free fall of a body under the action of gravity. And Galileo was anticipated, in this connection, by Domingo de Soto at the University of Salamanca [See W. A. Wallace, "The Enigma of Domingo de Soto: Uniformiter difformis and Falling Bodies in Late Medieval Physics", Isis 59 (1968): 384-401. Quoted by S.L. Jaki in "The Origin of Science and the Science of its Origin" (Regnery/Gateway. Inc.: South Bend, Indiana, 1979)].

Professor Hawking notes that Newton's "Principia", containing the theory of universal gravitation, lead to the idea of scientific determinism. Laplace pointed out that if, at one time, one knows the positions and velocities of all the particles in the universe, the laws of physics enable one to know the positions and velocities at any other time. Certainly, God does not interfere ordinarily with the laws of nature, but does this mean that human freedom is inexistent? As Arthur Holly Compton (1892-1962), the first to point out the wave-corpuscle character of light, forcefully points out, human freedom affirmation is based upon the inner conviction that moving one's own little finger at will carries far greater, and more immediate, evidence that all the deterministic laws of physics taken together [See "The Freedom of Man" (New Haven: Yale University Press, 1935), p. 26]. And what would be the use, in any case, of knowing exactly the positions and velocities of the 10^{80} particles in the universe? Its knowledge would not be sufficient in itself to predict the weather for tomorrow.

As Prof. Hawking remarked, Maxwell equations and Dirac's relativistic wave equations govern most of physics, chemistry and biology. But they are certainly unable to predict human behavior.

The Electro-weak theory and Quantum Chromo Dynamics constitute together the so called Standard Model, which aims to describe all physical interactions except gravity. But the Standard Model is clearly unsatisfactory, according to Hawking. The particles are grouped in an apparently unsatisfactory way. And the standard model depends on 24 numbers whose values cannot be deduced from first principles. It does not include gravity. It should be recognized that, after decades (at least 30 years) of intensive efforts, by the best theoretical physicists in the world, a consistent quantum gravity theory is not in sight. Professor Hawking says that, according to general relativity gravity is space and time. But, does he really mean that mass (inertial mass, gravitational mass) is only space and time, and nothing else?

Figure 1: Stephen Hawking (1942…).

The usual approach, according to Hawking, is to regard quantum space-time as a small perturbation of some background space-time. With gravity, the "effective coupling" is the energy or mass of the perturbation. In quantum theory, the electric fields, or the geometry of space-time, do not have definite values; they present quantum fluctuations which carry energy. According to Professor Hawking they have an infinite amount of energy, so that they fluctuate in all length scales from zero to infinity. But nobody has demonstrated yet, and it would be really difficult to demonstrate it observationally, that the universe is not finite.

Supergravity (1976), string theory (1985) and, somewhat later, M-theory, as a larger structure encompassing both, are briefly discussed by Professor Hawking in his lecture. The later could provide, according to him, a single theoretical formulation which would hopefully work in all situations.

Finally, he says that most people (including himself) did implicitly assume up to then that there is an ultimate theory eventually to be discovered. But this might not be the case: according to Professor Hawking himself, it may not be possible to formulate a theory of the universe in a finite number of statements. This is reminiscent, according to him, of Gödel's theorem, which says that any finite system of axioms is not sufficient to prove every result in mathematics, and hence in any physical-mathematical theory of the universe.

Gödel distinguished carefully between "mathematics" (statements, like 2+2 = 4) and "meta-mathematics" (statements about mathematics like "mathematics is consistent").

Then he shows that each mathematical formula can be given a <u>unique</u> number, the "Gödel number". He shows further that the meta-mathematical statement that the sequence of formulas A is a proof of formula B can be expressed as an arithmetical relation between the Gödel numbers for A and B. Therefore meta-mathematics can be mapped into arithmetic. Finally, he considers the self referencing Gödel statement G: "the statement G cannot be demonstrated from the axioms of mathematics". If G could be demonstrated, the axioms would be inconsistent, because one could demonstrate G and simultaneously show that it cannot be demonstrated. If G cannot be demonstrated, then G is true. As Professor

Hawking notes, by the mapping into numbers G corresponds to a true relation between numbers, but one that cannot be deduced from the axioms. Thus, mathematics is either inconsistent or incomplete: if it is consistent it is incomplete; if it is complete, it is inconsistent.

Professor Hawking notes that a physical theory is a mathematical model.

Of course, it is much more than a mathematical model, because it is about physical reality, not only about purely geometrical or algebraic concepts. A physical theory is self referencing, as in Gödel theorem. It should be therefore either inconsistent or incomplete. Not excluding, perhaps, both, in some cases.

Professor Hawking ends up his lecture saying that he used to belong to the camp of those who would consider themselves disappointed if there was not an ultimate theory, a theory which could be formulated in a finite number of principles, but he admitted that he had changed his mind.

A few years before Professor Hawking delivered his lecture on "Gödel and the end of Physics", Stanley Jaki, at the Universidad Autonoma of Madrid (UAM), spoke on the topic "There is such a thing as a last word in Physics?". A large number of students, mostly from the Faculty of Sciences, but also from other Faculties (Law, Economics, etc) was in attendance, probably in the Fall of 1992. A good number of undergraduate and graduate students, and many professors. Professor Jaki's talk was especially interesting, and many of those in attendance remember his talk many years later. Quite a few Law students present were really impressed to see that Physics could be so culturally relevant. Jaki left abundantly clear that, in spite of statements by some scientists of international reputation, the end of Physics was not in sight for two reasons: (1) Gödel's incompleteness theorems; (2) The fact that a discovery in the future of some subtle new physical interaction could not be discarded a priori.

CONFLICT OF INTEREST

The author(s) confirm that this chapter content has no conflicts of interest.

ACKNOWLEDGEMENT

Declared none.

REFERENCES

[1]   Hawing.S. "Gödel and the end of physics" (Cambridge University, July 20, 2002).

CHAPTER 5

Hawking on "The Grand Design", 2010

Julio A. Gonzalo[*]

Escuela Politécnica Superior, Universidad San Pablo CEU, Madrid

Abstract: After pointing out that Einstein's dream to discover the grand design of the universe was unrealistic, Hawkings and Mlodinov say that a few key developments (M-theory, COBE satellite's data, WMAP satellite's data) enabled physicists to achieve that dream. They conclude that God is not necessary because the universe is self-sufficient. The classical objections against God existence were known already to Greeks, Romans and Jews in Alexandria and had been rigorously reformulated already in the 13th century, Relevant quotes of St. Thomas Aquinus, Chesterton and S.L. Jaki are brought forward. It is the strongest mark of the divinity of man that he talks of this world a "strange world" though he has seen no other.

Keywords: Einstein, grand design of the universe, M-theory, COBE, WMAP, God's existence, St. Thomas Aquinus, Chesterton, Jaki, the infinite eccentricity of existence.

In his presentation [1] of "The Grand Design" (Stephen Hawking and Leonard Mlodinov), Professor Hawking says that in "A Brief History of Time" he left some important questions unanswered: Why is there a universe, Why is there something rather than nothing?, Why do we exist?, Why are the laws of nature what they are? Did the universe need a designer and creator?

After pointing out that Einstein's dream to discover the grand design of the universe or understanding the forces of nature was an unrealistic goal, he assures us that he himself, in "A Brief History of Time", was prevented from fulfilling Einstein's dream only because a few key advances had not yet been made: the development of M-theory, and some new observations by NASA's COBE and WMAP satellites.

COBE's observations (that the cosmic background radiation had a perfect Planck

*Address correspondence to Julio A. Gonzalo:** Escuela Politécnica Superior, Universidad San Pablo CEU, Madrid; Tel: 34-91-547-0815; UAN: 34-91-497-4767; Fax: 34-91-497-8579; E-mail: julio.gonzalo@uam.es

spectral distribution and that it had very small but detectable anisotropies of the order of one part in 10^5) and WMAP observations (that the time elapsed since the big bang was 13.7 ± 0.2 billion years; and that the present value of Hubble's parameter $H_O = R_O / R_O \approx 67 \pm 4\,\mathrm{km/s/Mpc}$) were certainly spectacular developments confirming the big bang model, a concept originated in George Lemaitre and developed by Gamov, Alpher and Herman, but were not confirmation of other post-Einstenian theoretical developments, in particular M-theory. Everybody knew f. i. that early small anisotropies were necessary in order to develop today's large cosmic anisotropies for the universe.

According to Professors Hawking and Mlodinov, "The Grand Design" <u>explains</u> why, by means of quantum theory, the cosmos "does not have just a single existence or history" but it has rather <u>every possible history simultaneously</u>.

A statement like this, <u>"untestable" observationally</u>, is easy to write down, but clearly does not mean anything physical.

To begin with, quantum theory is a statistical theory, and, therefore does not seem directly applicable to describe the collection of all possible universes, of which we can observe something, and physically measure something, only in one: <u>ours.</u>

In "<u>The Grand Design</u>", Hawking and Mlodinov do their best to <u>explain</u> how life evolved in the cosmos just as a result of a number of <u>extraordinary coincidences:</u> the earth is just at the right distance from a moderately hot and long lived star, occupying a favorable region of a standard galaxy, with just the right combination of chemical elements. Gonzalez and Richards do a much better job in "<u>The Privileged Planet</u>", without viewing in it a purely random set of coincidences. Of course, if in a lottery game you buy participations in all and every one of the numbers, you are likely to lose money. The cosmic <u>multiverse</u> amounts exactly to that.

In Professor Hawking words, <u>M-theory</u> is an explanation of the laws governing the <u>multiverse</u> and the only viable candidate for a complete "theory of everything".

At "A Brief History of Time" (1988) Hawking suggested that a Creator was unnecessary because he had arrived at a self consistent theory which avoided the need to specify any boundary conditions for the universe, therefore explaining, according to him, the apparent zero value of the cosmological constant, the apparent flatness of the universe, and why the universe was in thermal equilibrium at early times.

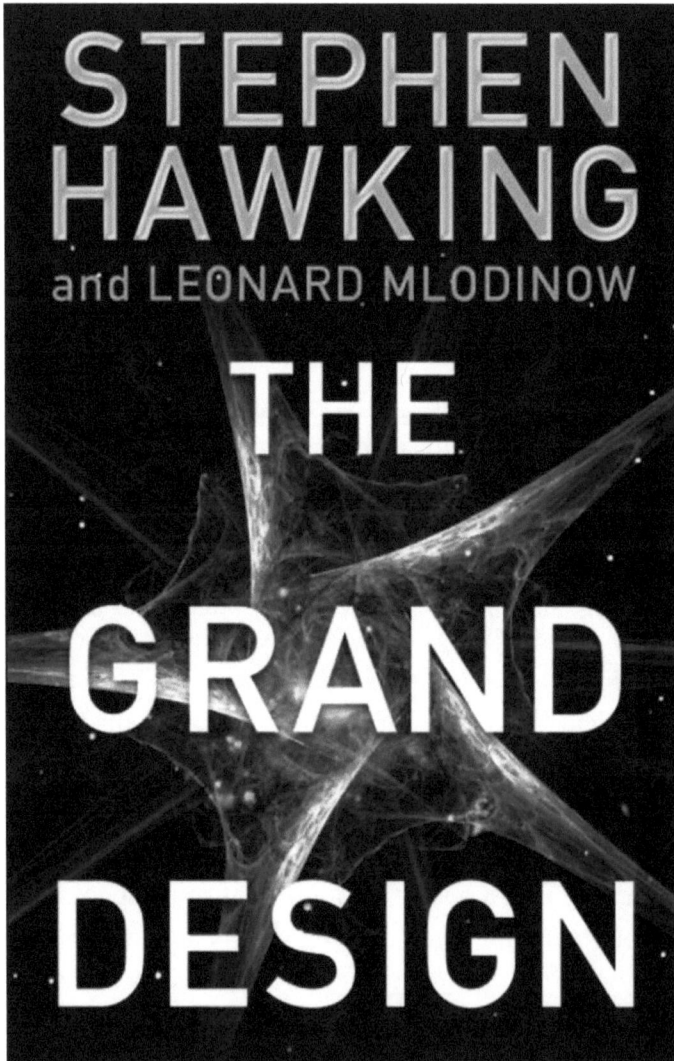

Figure 1: Cover of "The Grand Design" by Stephen Hawking and Leonard Mlodinow.

Then, in his Cambridge lecture on "Gödel and the end of physics" (2002) he finished it saying that he used to belong to the camp of those who would consider themselves disappointed if there was not and "ultimate theory" (a theory which could be formulated in a finite number of principles) but he admitted there and then that he had changed his mind.

And again, in "The Gran Design", Professor Hawking (or may be this time Professor Mlodinov) seemed to say, that he had <u>again</u> a candidate (the <u>only</u> candidate) for a <u>complete</u> "theory of everything".

Has he therefore changed his mind once more? We leave the reader to come to his own conclusion.

"The Grand Design" is presented as a succinct, startling and lavishly illustrated guide to discoveries that are altering our understanding and threatening "some of our most cherished belief systems"; as a book that will inform and "provoke" like no other.

In eight relatively brief chapters, which include standard popularizations of astronomical information, nuclear physics data and cosmological speculation, Hawking and Mlodinov offer very little new material.

The chapters in question cover: 1. The Mystery of Being, 2. The Rule of Law, 3. What is reality?, 4. Alternative Histories, 5. The Theory of Everything, 6. Closing Our Universe, 7. The Apparent Miracle; and finally, 8. The Grand Design.

Roger Penrose, the British mathematician who sheared in 1988 the Wolf Prize with Stephen Hawking, refers to Hawking in his book on "The way to reality", saying that he is not interested in arriving at a theory which corresponds to reality, because he does not know what it is. All he is interested in is in arriving at a theory which predicts the results of the measurements. Pensore position, on the contrary, is that the ontological question (what reality is in <u>itself</u>) has precedence in quantum mechanics and elsewhere. In any event, confusing the scientific question and the ontological question does not lead anywhere.

*   *   *

To prove a statement is to show that it can be put in terms of another statement that is clearer, and therefore more convincing. This other statement can be put itself in terms of another one that is clearer, and so on. But the process cannot go on forever. Regression to infinity leaves everything hanging on air. The same is true if, after a finite number of steps one comes back again to the initial statement- it would be circular reasoning- leaving also everything hanging on air. So, the process must end in a statement that is sufficiently <u>convincing</u> or <u>evident</u> to justify the initial statement. <u>A first step</u> is always necessary. If there is no first step, things remain hanging in air for ever. That is why a "nude negation" is useless as a starting point to explain anything. One must start always accepting something as valid to build up from there.

Professor Hawking holds that God is not necessary because the universe is self-sufficient to justify itself.

Let us analyze this statement from the logical and the ontological point of view.

Professor Hawkings argument to deny God's existence is certainly <u>not new</u>. It dates back at least to the 13th century, and probably back to much earlier times. The times during which the Greek, Roman and Hebrew cultures meet at Alexandria.

A <u>classical objection</u> against God's existence can be summarized as follows: It appears that everything we see in the world can be accounted for by two principles: all natural things by nature; all voluntary things by human will and reason. Therefore there is no need to assume God's existence.

The <u>classical reply</u> to this objection is: when Nature works for a determinate end Nature does it under the direction of a higher agent. Whatever is done voluntary can be traced back to some higher cause. All changes in things that are changeable and contingent can be traced back to an immovable and necessary first principle: God.

[For a more complete discussion, see f.i. Peter Kreeft, "Summa of the Summa"- [2] The Essential Philosophical Passages of St. Thomas Aquinas <u>Summa Theologica</u> Edited and Explained for Beginners (Ignatius Press: San Francisco, 1990)].

Some scientific popularizers, when confronting the vastness of space filled with billions of galaxies made up of billions of stars, arrive to the conclusion that God is not likely to have wasted so much effort to end up with only a little planet, the Earth, inhabited by men. But, for God's omnipotence, of course, creating a universe like ours is like nothing (Truly, much less than "a free lunch").

It is obviously very difficult for us (not to say impossible) to understand the creation of free beings, be they men or angels, because we, being free, (but, with a limited intellect and a limited will), are utterly unable to create anything endowed with freedom.

As Stanley Jaki says in his book on Chesterton [3]:

> "If the universe was the chief bone of contention or, if one prefers, the great continental divide between heretics and orthodoxy, then in a Chestertonian discussion of orthodoxy the universe could be expected to loom large. Indeed it did. The presentation of the <u>universe</u> as a cosmic town both <u>queer</u>, that is, so specific as to border on queerness, and very <u>homey</u>, that is something very familiar and inviting on mere sight, was the objective of <u>Orthodoxy</u> set by its author". (Emphasis added)

The alternatives are <u>solipsism,</u> which reduces the whole cosmos to a nullity (as Chesterton says "If the cosmos is unreal there is nothing to think about it"), or <u>pantheism,</u> with a choice between pessimism and optimism (Chesterton: "The transcendence and distinctness of the deity which some Christians now want to remove from Christianity, was really the only reason why any one wonted to be Christian. It was the whole point of the Christian answer to the unhappy <u>pessimist</u> and the still unhappy <u>optimist</u>". Emphasis added).

Stanley Jaki summarizes the greatest story in the whole history of science, the history of the development of scientific cosmology in the 20th century, in the following words:

> "As the <u>Einstenian approach</u> unfolded more and more specifics about the universe, <u>those specifics</u> reveled more and more their <u>mutual coherence</u>. First came the discovery, theoretical and empirical, of the expansion of the universe. There followed the connecting of the genesis of the elements with the genesis of the universe. Scientific minds tired of, or simply not

tuned to the stark specificity of things around them, are now confronted with a cosmic specificity which defies imagination, and are seized by a gripping sensitivity for metaphysics. The result was prophetically portried in Chesterton's final explanation of the optimism of Dickens world view, tragic as the world would appear in his novels. From the remark that the ultimate basis of that optimism ("cheerfulness in Chesterton words) was Dickens' perception of the strangeness of every person and situation".

In Chesterton own words:

"But when all is said, as I have remarked before, the chief fountain on Dickens of what I have called cheerfulness, and some prefer to call optimism, is something deeper than a verbal philosophy. It is after all an incomparable hunger and pleasure for the vitality and the variety, for the infinite eccentricity of existence. And this word 'eccentricity' brings us perhaps, near to the matter than any other. It is perhaps, the strongest mark of the divinity of man that he talks of this world as 'strange world', though he has seen no other…"

For a full discussion of Professor Jaki comments on Chesterton as champion of the universe see "Chesterton, A Seer of Science" (University of Illinois Press: Urbana and Chicago, 1986).

CONFLICT OF INTEREST

The author(s) confirm that this chapter content has no conflicts of interest.

ACKNOWLEDGEMENT

Declared none.

REFERENCES

[1]   Hawing. S, Mlodinov. L. "The Grand Design" (Bantam Books: New York, 2010).
[2]   Kreft. P, "Summa of the Summa" (Ignatius: S. Francisco, 1990).
[3]   Jaki S.L,"Chesterton, A Seer of Science" (University of Illinois Press: Urbana and Chicago, 1986).

CHAPTER 6

Hawking and the Universe

Manuel M. Carreira S.J.

Universidad Pontificia de Comillas, Madrid

Abstract: First, the limits of Science, Philosophy and Theology, and the proper methodologies of each field of knowledge must he recalled in order to speak meaningfully of material reality, living or non-living. The existence of infinite universes as the theoretical reason why Einstein's cosmological constant is so close to zero is impossible to verify for those other universes. Eddington, Dicke, Carter, Barrow, Wheeler and Hawking himself have underlined the need for most precise values of the different cosmic parameters (Anthropic Principle). Finality is a metaphysical problem when we speak of the Universe and to deny it leaves us with an absurd. We should admit that there is no scientific answer as yet for any of the important questions posed by biology. "The Emperor's New Mind" of artificial intelligence is a fraud, as pointed out by Penrose. In "The Great Design", their authors present the multiplicity of undetectable universes (than appear out of "nothing") as the explanation of the fact that the one we detect is suitable for life. No Science can predict what I will do next minute. Neither can it predict the free activity of the Creator who holds the Universe in existence.

Keywords: Science, philosophy, theology, Einstein, Eddington, Dicke, Carter, Barrow, Wheeler, Hawking, Penrose, "The Grand Design", multiple universes, the Creator's free activity.

Within the traditional themes [1] where Science, Philosophy and Theology meet, not always amicably, the questions regarding the origin of material reality, living or non-living, are the most frequently debated. The contributions of people who are regarded as experts seem to give rise to sensational headlines, not rarely distorting the meaning or the real worth of different pronouncements. It becomes necessary, time and again, to recall the limits and methodologies proper to each field of knowledge, in order to establish the value of each view and of the arguments that are presented to support it. Being famous in any sense of the word does not imply an infallibility or even basic expertise outside of a very specialized field.

*Address correspondence to Manuel M. Carreira: Universidad Pontificia de Comillas, Madrid; Tel 34-91-540-6101; Fax 34-91-372-0218; E-mail: ecarreira@res.upco.es

A second book, Black Holes and Baby Universes, suggested the existence of infinite universes as the theoretical reason why the possible cosmological constant proposed by Einstein has a value extremely close to zero. But all those other universes are impossible to verify, even by their very concept, and as said before, whatever in principle cannot be experimentally checked is not accepted as science. With Hawking's own words, "to speak of the baby universes might be considered as much a waste of time as discussing how many angels can fit on a pinhead": A quite correct evaluation of something that is -at best- science fiction and does not deserve serious attention [2].

Intimately related to the reason for existence -of the only Universe we can study- is the question of the initial adjustment of its properties. Beginning with Eddington, Dicke, Carter, Barrow, Wheeler, and Hawking himself, have underlined the need for a most precise value of the different parameters -cosmic density, the strength of the four forces, particle masses- in order to provide conditions for the existence of intelligent life, at least in our planet (the Anthropic Pinciple). The physical world and its evolution appear as fine tuned to reach this end, and the alternative explanations can be reduced to two: either we admit an all-powerful and intelligent Creator or we fall again into the unscientific infinitude of parallel or successive unobservable universes where the full variety of properties will be found. Their only justification would be the need to provide "by chance" at least one that is correctly suited for our existence, but without giving a reason for the existence of any of them, except to satisfy some mathematical suppositions.

It is also the way that the existence of life, of conscious activity, of intelligence - and of science itself- is supposed to be justified: whatever is possible a priori must de facto exist We could apply the rigorous test of simplicity -Ockham's Razor- and ask what is simpler: a single well-made Universe or an infinite pile of worthless and sterile ones to allow one to be correct by chance. Or one might fall into the ultimate vicious circle as proposed by Wheeler ("The Universe as Home for Man", 1978): only what is observed is real, and Man as the observer makes the Universe real, obliging it to begin with the correct parameters to make Man possible, so that the Universe itself might be possible!

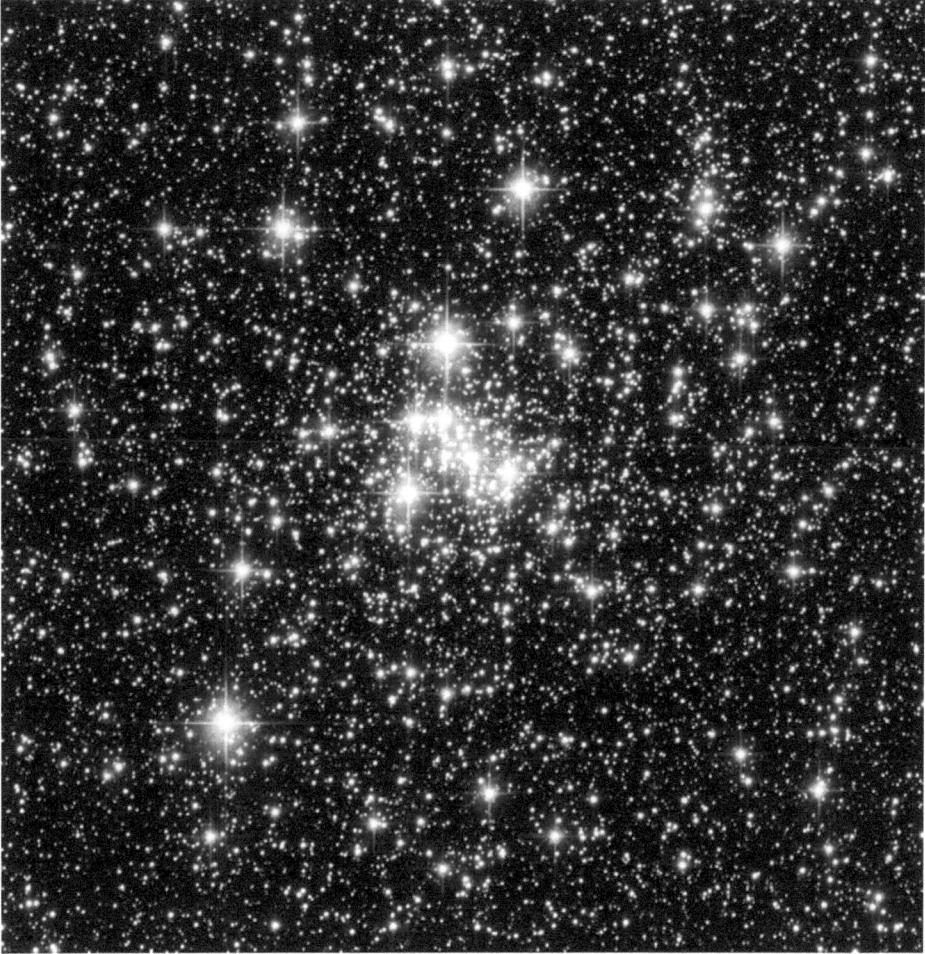

Figure 1: Olbers Paradox.

If finality is an obvious metaphysical problem when we speak of the Universe and of the Anthropic Principle, and to deny it leaves us with an absurd, it again appears as unavoidable when talking about life and its origin Even Monod ("Chance and Necessity") has to admit some "teleonomy" in a living being: the most obvious property -even of the simplest cell- is that it has a unity of activity directed to its self-preservation, something that "chance" cannot explain, because chance is not a physical force and it can never be a sufficient reason for order. We should admit, without committing any injustice, that there is no scientific answer as yet for any of the important questions posed by biology, even if general statements are made that sound more or less plausible.

We don't know where, when or how life appeared on Earth, and 50 years after Urey and Miller synthesized some amino acids in the laboratory (with a series of electrical sparks in a "primitive atmosphere") we are quite far from getting anything as complex as a molecule of DNA or RNA. We cannot convincingly explain how evolution has managed to develop -by random genetic mutations- very specialized organs as found even in insects or arthropods. And no organic evolution or adaptation can explain abstract thought, true intelligence.

One cannot establish any convincing evolutionary argument to show that there was a real need for intelligence to develop as a survival element: very primitive species, from bacteria to coelacanths and ants, have survived during geological periods spanning hundreds or even thousands of millions of years. Nobody can pinpoint an ecological environment whose dominion would be the stimulus for the surging of conscience, language, the search for Truth, Beauty and Good, the traits that describe us as human and that have no value of adaptation for survival in any specifiable physical milieu.

While different species undergo changes that make them more suited to a given habitat, Man has no other specialization than the mysterious thirst for knowledge and harmony that appears in the arts, in philosophy, in mathematics and science: a series of "products" that have no measurable physical parameters, no mass, temperature or electrical charge The only logical consequence of this fact will be to attribute something that doesn't fit the operational definition of matter to a non-material source, such that it cannot be explained by simple material evolution from previous stages.

It is necessary to say again and again that the science of matter only admits four interactions -forces- to give an explanation of whatever happens in our surroundings, be it in the physics of inanimate matter or at the level of biology. We have a clear description of the way those forces act, in terms of attractions, repulsions, wave or particle exchanges: nothing appears in those activities that might point to the possibility -still less, that would give a sufficient explanatory reason- for consciousness and abstract thought. To consider that a genial intuition is the product of chemical or electrical signals between neurons would be as senseless as expecting that a great literary work will by chance appear as the end result of electrical currents in the transistors of an electronic circuit. "The Emperor's New Mind" of artificial intelligence is still a fraud, and that will always be so, as Penrose writes in the book by that title.

Finally, in this hunt for answers with so few data available, we cannot expect that we will seriously be able to establish the probability of extra-terrestrial life, especially intelligent life The famous Drake equation is at best a plausible exercise in pure guessing and its final predictions are only the reflection of imagination, pessimism or optimism, when giving values for the different variables, even if it has the merit of pointing out the multiple factors to be taken into account.

Any final result of the computation is not science as long as we can't say how life originated here, and the common assumption that it would do the same in any environment where a small pond contained the necessary ingredients and chemical conditions at the correct temperature, still needs a minimal experimental confirmation. One could expect that a flask of pure distilled water, enriched with the well mixed contents of two fresh chicken eggs, would spontaneously produce at least one living cell in the laboratory: after all, two chicks with millions of different cells would build themselves up from that material if put in an incubator. But nothing like that has ever happened. The only undeniable fact is the marvelous existence of full panoply of life in our privileged planet, in the last seconds of the "cosmic year" since the Big Bang, with the crowning achievement of life, intelligent Man.

STEPHEN HAWKING'S "THE GREAT DESIGN"

A book by Hawking with that title has recently been the cause of headlines in the mass media. It has 8 chapters, the central 6 giving a readable description of the science of the last century with nothing new. Still, with an almost obsessive insistence, they present as certain the supposed multiplicity of undetectable universes (that appear from NOTHING) that is meant to explain the fact that the only one we can know is suitable for life. It is also implied that unification theories (Superstrings and the M Theory) are the definitive answer in Physics, even if they lack the most minimal experimental confirmation and after 30 years they still cannot be presented in a unique coherent form (see the book by LEE SMOLIN, The Trouble with Physics).

We should state before anything else that the concept of matter in present science is defined in terms of its interactions, the two long-range forces of gravity and electromagnetism and the two short-range of strong and weak nuclear interactions at the nuclear level. Matter is the totality of particles, energy, the physical vacuum, space and time. We can detect the activity of particle-antiparticle formation and destruction in the physical vacuum, and a gravitational field distorts empty space and changes the rate of phenomena that indicate the passage of time. This is the way modern science speaks and there are experimental proofs of such activities at every level.

But the section of the book singled out for underlining by lay commentators (not by scientists) is the final chapter where very briefly -in half a page- it states the rather bizarre idea that from true NOTHING a Universe can be formed spontaneously by the force of Gravity. We have a play of words that is plainly absurd: in nothingness (NOT THE PHYSICAL VACUUM) we have no properties, therefore no gravity: only existing matter in some form can produce gravity. The most basic law of physics is precisely that no physical process ever creates or annihilates anything, but only changes it. No calculation can be performed except stating initial conditions that can explain a development of the existing matter according to its properties: if the initial state is the total absence of matter, nothing will develop. And we cannot say that the present Universe is "Nothing" even if we calculate an equal amount of positive and negative energy in it. With a simple example: If I have particles with electrical charges of both signs in equal quantities I still have electrical charges and I would never say that I have nothing.

Another nonsensical statement appears at the beginning of the book, in the first chapter, where with an a priori materialistic assurance it is asserted -as outside any discussion- that the only thing that exists is matter, with fixed laws. As a consequence, human free will must be an illusion, even if it is the most obvious personal experience of each one of us. I am sure that Hawking - or his lawyers- would not accept my excuse if I were to tell him that I am physically predetermined not to repay an amount of money owed to him. In physics, it seems contradictory to later state -talking about Quantum Mechanics- that the world is determined to act in such a way that everything is ruled by chance: if there is a constancy (at least in the frequency of specific results), there must be a sufficient reason for it, that cannot be a childish "just because" disguised as "chance" in probabilistic computations. Chance is not a measurable force and it cannot be detected by any instrument, nor can it enter with a specific value in an equation. It really has a meaning only to deny a logical connection of unrelated events, that we observe happening with an unforeseeable coincidence of time and place.

As a consequence of his refusal to accept any reality superior to matter, Hawking denies the possibility of a non-material Creator, as well as any miraculous intervention, no matter how well historical miracles might be established as facts. This implies either a really unbelievable ignorance of history or the refusal to accept whatever doesn't fit personal prejudices. But to the argument that science would be impossible otherwise -it must be able to predict the future with certainty- one can reply with our own human behavior. No science can predict what I will do in the next minute: keep in my hand a pencil or lay it down on the table, but science does not crumble because of that. Neither does it become impossible by not predicting the free activity of the Creator who holds the Universe in existence, an activity that is not arbitrary or meant to confuse scientists, but exercised for superior reasons, rarely, and only in a religious context.

When asked for the reason for the second popular book, "Baby Universes and Black Holes ", Hawking replied -with a sense of humor- that he "needed the money". Perhaps he needed it again, or at least he wanted to impact the public with shocking statements that add nothing to science. His original idea of many years ago - Hawking's Radiation allowing the evaporation of black holes- has not been followed by comparable insights since. It is admirable that he still has the strength to do any scientific work in spite of his crushing health condition, but I can honestly think that with normal health he would not draw the attention of the mass media.

CONFLICT OF INTEREST

The author(s) confirm that this chapter content has no conflicts of interest.

ACKNOWLEDGEMENT

Declared none.

REFERENCES

[1]   Carreira M. M. S.J., "El hombre en el cosmos" (Ciencia y Cultura: Madrid, 2009).
[2]   Hawing. S, Mlodinov. L "The Grand Design" (Bantam Books: New York, 2010).

CHAPTER 7

The Origin of Science in the Christian West

Julio A. Gonzalo[*]

Escuela Politécnica Superior, Universidad San Pablo CEU, Madrid

Abstract: As Whitehead pointed out, science sprang from the heart of the Christian West. This has been well documented by P. Duhem and S.L. Jaki. This assertion is illustrated by quotes from various Christian natural philosophers precursors of Copernicus, Galileo, and Newton, including Abelard of Bath, Tierry of Chartres, Robert Grossetteste, William of Auvergne, Thomas Aquinas, Roger Bacon, Etienne Tempier, Jean Buridan and Nicole Oresme. Christian belief provided a "cultural matrix" for the growth of science.

Keywords: Whitehead, origin of science, P. Duhem, S.L. Jaki, Christian belief as "cultural matrix" for science.

Science became a self-sustaining enterprise only in the Christian West:

"... as Whitehead pointed out, it is no coincidence that science sprang, not from Jonian metaphysics, not from Brahmin-Buddhist-Taoist East, but from the heart of the Christian West..." (Walker Percy, "Lost in the Cosmos")

[Quoted in www.columbia.edu/cu/angustine/a/science_origin.html]

The simplistic popular notion that science is hostile to religion, fueled by not a few well written popular scientific books (those of Sagan, Asimov, etc) is simply not true. Exactly the contrary: modern science had its origin in medieval Christian Europe, as meticulously documented by the great theoretical physicist and historian of medieval science Pierre Duhem (1861-1916) in his monumental work "Le systeme du monde" (A. Hermans et Fils: Paris, 1913...), and further expounded and developed in numerous books and publications by Stanley L. Jaki (1924-2009), one of the foremost historians of science of the twentieth century.

*Address correspondence to Julio A. Gonzalo: Escuela Politécnica Superior, Universidad San Pablo CEU, Madrid; Tel: 34-91-547-0815; UAN: 34-91-497-4767; Fax: 34-91-497-8579; E-mail: julio.gonzalo@uam.es

As Jaki recounts in detail in "The Origin of Science and the Science of Its Origin", some enlightened European freethinkers contemporary of Voltaire, after reading extensive reports on China by Jesuit missionaries, were impressed by Chinese achievements in ethics and moral philosophy, architecture, engineering, arts and crafts, but very little in science. The work of Fr. Louis Lecompte SJ, "Nouveaux Memoires sur l'état present de la Chine" (1696) was composed of fourteen long letters to various civil and ecclesiastic French dignitaries. It was soon translated into English, German and Dutch, and it covered many topics on Chinese geography, politics, history, literature, arts and crafts, and also science. The science that Fr. Lecompte had in mind to make a comparison with Chinese science was Euclid's and Ptolemy's science. However, at the time, European science had just achieved full maturity in the "Principia Mathematica" of Newton. A maturity with roots at medieval Christendom, in the seminal work of Jean Buridan and Nicole Oresme, whose pioneering ideas provided fertile ground for the decisive developments of Copernicus, Galileo and Kepler, and, finally, Newton.

In view of the considerable talents abundantly demonstrated by the Chinese through their millenary history, their achievements in science were certainly meager, compared with those in Europe at the time.

Then the question: Why not in China?

But the proper question to be asked was not the one made by those European free thinkers: That question should have been: Why science, a self-sustaining science, worth of that name, had developed only in one cultural matrix, the cultural matrix of Medieval Christendom?

Other great civilizations in world history could be justly proud of their stupendous achievements in architecture, public works, arts and crafts, drama, literature, even in philosophy and logic. But not in science proper. At least in any degree comparable to the level achieved at the beginning of the eighteenth century in Europe. An achievement with roots in Medieval Christendom as shown by Duhem and Jaki.

In the period going from the early twelfth century to the time of Buridan and Oresme, at which it was introduced for the first time the concept of "impetus" and the concomitant idea of "inertial motion", one can see developments which lead directly to the formulation of the fundamental laws of motion (Newton's laws), developments connecting in one stroke motion here on Earth and motion in the sidereal realm: the motion of the "planets" or "vagabunds".

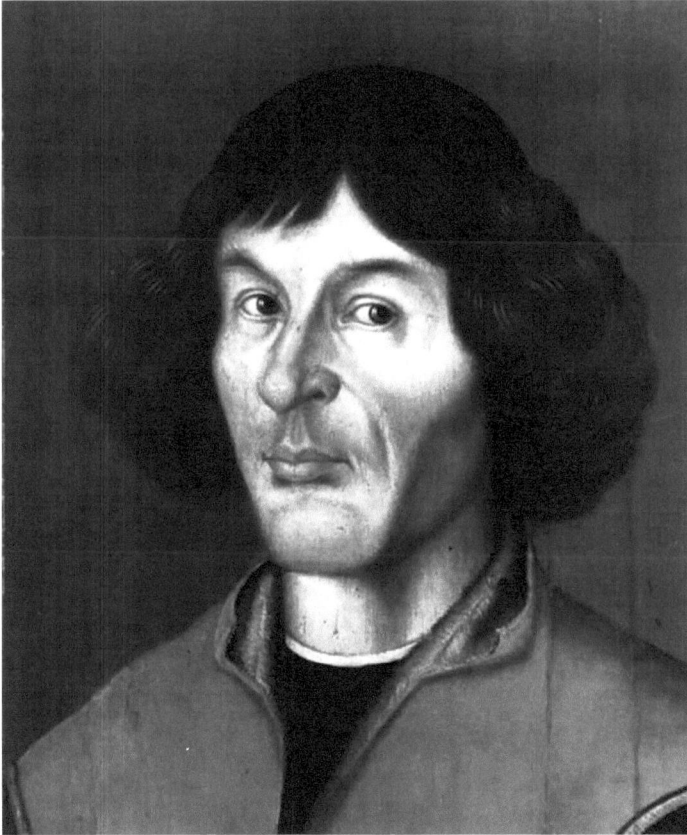

Figure 1: Nicolaus Copernicus (1473-1543).

Slowly at first, then at a fast pace, Copernicus, Galileo and Kepler, prepared the way to Newton [1].

Abelard of Bath (c. 1125) embarks on long and arduous journeys in the quest of learning, going as far as the Middle East, and brings to medieval Europe the elements of trigonometry, the art of making astrolabes and Euclid's geometry. His contacts with the Muslim learned men made him aware of the ongoing struggle in the Arab culture to reconcile faith and reason. Abelard is on record remarking to his nephew that many of his contemporaries, including Muslim and Jews men of learning, identified God with Nature. There was, at this time and at any other time, a strong tendency to identify nature with the ultimate entity. A tendency to be overcome only if men are willing to recognize their dependence on a truly transcendent Creator. Abelard favors, whenever possible, natural explanations over miraculous ones, showing his healthy proclivity to a true scientific attitude when facing the physical world: "I do not detract from God… Whatever there is it is from Him and through Him. But the realm of being is not a confused one… Only when reason totally fails should the explanation of the matter be referred (directly) to God".

In other words, Abelard sees in Nature Nature's God without any need to denying the world of the supernatural, which, if real, is also God's. Medieval men were tempted too by the mirages of fatalism and astrological pantheism.

Thierry of Chartres (d.c. 1155) rises well above the Greek animism and pantheism, latent even in the best literary exponents such as Plato's "Timeus", saying: "Moses intention was to show that the creation of all things and the formation of men was made by the only one God to whom alone worship is due. The usefulness of [Mose's] works is the acquisition of knowledge about God through His handiwork". For Thierry "there are four kinds of reasons that lead man to the recognition of his Creator: the proofs are taken from arithmetic, music [harmony], geometry and astronomy". If the Creator has actually arranged everything "according to number, measure and weight", as recorded in the Book of Wisdom, man's intellectual understanding of the world has to have a mathematical, scientific character.

Robert Grossetteste (c. 1168 – 1253), possibly the first chancellor of Oxford University, he was even more explicit about the mathematical understanding of nature: "The usefulness of considering lines, angles and figures is the greastest, because it is impossible to understand natural philosophy without them. They are efficacious throughout the universe and its parts and the properties related (to them), such as rectilinear and linear motion". Grosseteste's investigation on the rainbow is a good example of his scientific methodology, which include seminal programs of induction, falsification and verification. He rightly attributes the rainbow to light's refraction rather than to its reflection, as done before incorrectly by Aristotle and Seneca. Grossetteste's methodology depends on the idea of the Creator as a wholly rational and personal Planner, Builder and Maintainer of the Universe.

William of Auvergne (d.c. 1249) in "De Universo" makes fragmentary references to magical and astrological aphorisms but he does not succumb to the irrational in his quest for understanding, and disputes continuously with "Manichaeism, fatalism, pantheism, star worship" and similar betrayals of man's rationality, which he associates with the Greeks of old, with the Saracens and with the hermetic philosophers. In his lengthy discussion of the Great Year, identified with the period of 26,000 years needed to complete a precession of the equinoxes, William of Auvergue is aware of the belief in "eternal returns", which he rightly sees as the embodiment of a pagan, non-Christian world view. Defending right reason against the Mutakallimun, (devout and learned Muslim philosophers), he points out the fatal error of not distinguishing clearly between primary and secondary causality.

Thomas Aquinas (1225 – 74) embarks in a giantic effort to bring together <u>faith</u> and <u>reason</u> "in a stable synthesis". His theory of knowledge is moderate realism and the doctrine of the analogy of being becomes the key of his metaphysics. His resolute commitment to give reason its due meant a generous acceptance of the Aristoteliam system, then for almost two millennia the epitome of a rational explanation of the world. It was motivated (according to Stanley Jaki) by his attention to contemporary Muslim theologians and philosophers. His "Summa contra gentiles" is intended to counter the occasionalism and the fatalism contending with each other within the Muslim contemporary culture. His "Summa Theologica", a work of synthesis, aims to show that the reason for the existence of the cosmos is its subordination to man's supernatural destiny. Aquinas, contrary to his master Albertus Magnus, is not interested in experimental investigation, but both, disciple and master, agree in the all important point of rejecting the inevitability of "eternal recurrences" in the world. In his "De fato", Albertus is most explicit about the history of the question of "eternal returns" in Plato, Aristotle, the Sotics, Ptolemy, the Arabs (Albumasar especially), all of them inclined in favor of "eternal recurrences", and in the early Church Fathers, who were opposed.

In his famous Five Ways, Aquinas, beginning with specific characteristics of actual existing beings, arrives to the necessary existence of a Prime Mover, an Uncaused Cause, a Necessary Being, a Most Perfect Being and an Ordering Intellect, who everyone understands to be God.

It may be noted that, towards the end of the eighteenth century, enlightened philosophers began to take as an indisputable fact that there is no way to go from the "cosmos" ("a bastard product of the metaphysical cravings of man" according to Kant) to the Creator. Thanks to Einstein's General Theory of Relativity, we have now a contradiction-free concept of the "cosmos", a finite "cosmos", which is a perfectly valid ground to go from its existence to the Creator's existence.

Let us quote Jaki on Chesterton to that effect:

> *"No wonder that (Chesterton)... spoke devastating words of that philosopher, Kant, who more than any other succeeded in leading mankind into the belief that the universe was the bastard product of metaphysical cravings: Long essays on Kant and the German idealists contain far less than these few words of Chesterton: 'The note of our age is a note of interrogation. And the final point is so plain; no skeptical philosopher can ask any question that may not equally be asked by tired child on a hot afternoon. Am I a boy?-Why am I a boy?-Why aren't I a chair?-What is a chair? A child will sometimes ask questions of this sort for two hours. And the philosophers of Protestant Europe have asked them for two hundred years.' ..." (See Stanley L. Jaki, "Chesterton, a Seer of Science", University of Illinois Press Urbana and Chicago, 1986).*

Roger Bacon (1214-94). It seems that Bacon's impetuous efforts to secure the service of science for the Christian faith lead him to compose his "Opus majus", which resulted in his temporary imprisonment. Apparently, his novel views stressing too much the inexorable determinism of events in nature made them, at one time, suspicious of being incompatible with man's freedom and moral responsibility. But, according to Jaki, Bacon did not capitulate, as most Arab commentators of Aristotle, with the idea of cyclic determinism, and he made a clear distinction between man's supernatural destiny and his earthly every day existence, between final (primary) causes and efficient (secondary) causes. For him man's knowledge about nature was always partial, not "a priori" in character. One of the most learned and celebrated teachers at Oxford in his time, he was a scientific pioneer in controlled experiment and accurate observation of natural phenomena. He said that mathematics was the gate way to science and experience, or verification, the only basis of certainly.

Etienne Tempier (bishop of Paris), in 1277, condemned 219 propositions mostly against Siger of Brabant and his followers, excluding the eternity of the world (p. 83-91) and the perennial recurrence of everything every 26000 years (p. 92). He was affirming in substance the rigorous contingency of the world with respect to a transcendental Creator, source of all rationality on heaven as well as on Earth. Pierre Duhem, the great French physicist and historian of medieval physics, takes the decrees of Bishop Tempier as the starting point of a new era in scientific thinking. In the decrees, the possibility of several worlds was recognized in principle (p. 27); it was rejected, on the other hand, the existence of animated, incorruptible and eternal superlunary bodies (p. 31-32); it was admitted the possibility of rectilinear motion for celestial bodies as part of a celestial machinery (p. 75); it was rejected the deterministic influence of the celestial stars on the lifes of individual men from the instant of their birth (p. 105); it was rejected the provenance of a "first matter" from celestial matter (p. 107), *etc.* All of it aimed at defending the freedom and the exclusive rights of the Creator when he created heaven and Earth.

The bishop's statements were not binding in the universal Church but were intended at upholding orthodoxy in the University of Paris where Siger Brabant had been teaching then for more than a decade.

As pointed out by Alfred North White head in 1925 in his "Lowell Lectures":

> *"I do not think however that I have even yet brought up the greatest contribution to the formation of the scientific movement. I mean the*

inexpugnable belief that every detailed occurrence can be correlated with its antecedents in a perfectly definite manner, exemplifying <u>general principles</u> (emphasis added). Without this belief the incredible labours of scientists would be without hope. It is this instructive conviction, vividly posed before the imagination,which is the motive power of research: that there is a secret, a secret which can be unveiled. How has this conviction been so vividly implanted in the European mind?"

Jean Buridan (1300-58). He, according to Stanley Jaki, through his commentary on "De Caelo" by Aristotles, had an unmistakable influence on Galileo. His "Questiones super quattor libris de caelo et mundo", through the slightly modified version by his disciple Albert of Saxony, were commonly available at medieval European universities from Salamanca to Crakow. The Aristotelian clear-cut distinction between superlunary and sublunary matter were dealt a decisive blow in Buridan's work. He reminded his readers that the heaven could decay and that the Creator was perfectly free to amihilate the world if He so wished: "In natural philosophy one should consider processes and causal relationships as if they always come about in a natural fashion; God is no less the cause, therefore, if this world and its order have an end, than if this world is eternal". Buridan did not accept yet the rotation of the Earth (his disciple Oresme later did) but, against Aristotles, he held that continuous, thought relatively small changes between the Earth and the fixed stars were possible.

Buridan proposed, against the mistaken Aristotelian notions on motion, his new concept of "<u>impetus</u>", a quality implanted in the moving body by the mover, and his notion of "<u>gravity</u>", a property innate to all massive bodies. And after reviewing the usefulness of his new theories to describe various motions here on Earth, he dared to outline their usefulness for celestial mechanics. He wrote, in the same breath, about a jump and about ordinary planetary motion, with the Creator as the ultimate agent imparting a given quantity of motion to the various parts of the universe: "He (God) created the world, moved all the celestial orbs as He pleased, and in moving them He impressed in them "impetuses" which moved them without his having to move them any more except by the ... general influence whereby He concurs as co-agent in all things which take place... And these impetuses which He impressed in celestial bodies were not decreased not corrupted afterwards because there was no inclination [to it]... Nor there was resistance which would be corruptive or resistive of that impetus... But I do not say this assertively, but [rather] tentatively..." [For a more complete discussion see Stanley L. Jaki, "Science and Creation: From eternal cycles to an oscillating universe" (Lanham: New York, 1990) and references therein].

Nicole Oresme (1323 – 82). The most outstanding disciple of Buridan, he wrote with great originality on a variety of topics, including monetary theory, astronomy and algebra. His commentary on Aristotle's "De Caelo" is considered today a classic of early scientific literature. For Oresme the perfection of the laws of nature is but a modest reflection of the infinitely perfect attributes of the Creator. Oresme allowed corruptibility in the celestial motions of celestial bodies only in the restricted sense that these motions were frictionless. Against the possibility of eternal recurrences, he noted that the periods of the planets are most likely incommensurable. He pointed out that, in any case, such coincidences should occur after periods much larger than 26000 years, or the period of the precession of the equinoxes (the length assigned in ancient Greece to the "Great Year"). Oresme parted with Aristotelian necesitarianism, which implied that several worlds would require several Gods. He responded: "One God governs all such worlds".

Many years later, "impetus" would be correctly redefined as "momentum", *i.e.* the product of inertial mass times the velocity imparted to the moving body. The full process from Buridam and Oresme to Newton took three hundred years. Copernicus (1473 – 1543), with the "impetus" theory behind him, gave the crucial step of postulating the heliocentric description of planetary motion. Galileo (1564 – 1642) and Kepler (1571 – 1630) did the next important steps. A Christian worldview was essential in the whole process.

HOW DID CHRISTIAN BELIEF PROVIDE A "CULTURAL MATRIX" FOR THE GROWTH OF SCIENCE?

Stanley Jaki in his booklet on "Christ and Science" [2] gives four reasons:

- "... the Christian belief in the Creator allowed a breakthrough in thinking about nature. Only a transcendental Creator could be thought of as being powerful enough to create a <u>nature</u> with <u>autonomous laws</u> without his power over nature being thereby diminished. Once the basic among those laws were formulated, <u>science</u> could develop on its own terms".

- "The Christian idea of <u>creation</u> made still another crucially important contribution the future of science. It consisted in putting <u>all material beings on the same level</u> as being mere creatures… The assumption would have been a sacrilege in the eyes of any one in the Greek pantheistic tradition or in any similar tradition in any of the ancient cultures".

- Finally, <u>man</u> figured in the Christian dogma of creation as being specially created in the <u>image of God</u>. This image consists in man's rationality as somehow <u>sharing in God's own rationality</u> and in man's condition as an ethical being with eternal responsibility for his actions. "Man's reflection on his own rationality had therefore to give him <u>confidence</u> that his created mind <u>could fathom</u> the rationality of the <u>created realm</u>".

- "At the same time, that very createdness could caution man to guard against the ever-present temptation <u>to dictate to nature what it ought to be</u>. The eventual rise of the experimental method owes much to that Christian matrix".

In contrast to those Christian natural philosophers, who brought up the birth of science in Medieval Europe, some of today´s leading theoretical physicists tray to dictate to nature what it ought to be.

CONFLICT OF INTEREST

The author(s) confirm that this chapter content has no conflicts of interest.

ACKNOWLEDGEMENT

Declared none.

REFERENCES

[1]   Jaki S.L., "Science and Creation" (University Press of America: Lanham, MD, 1990).
[2]   Jaki S.L., "Christ and Science" (Real View Books: Royal Oak, Michigan 2000).

CHAPTER 8

Science: Western or What?

Julio A. Gonzalo[*]

Escuela Politécnica Superior, Universidad San Pablo CEU, Madrid

Abstract: Modern science is characterized by an impressive capability to describe in quantitative terms an enormous variety of natural facts. If the world had not been made rationally, scientific knowledge would be impossible. P. Duhem in his "Le systeme du monde…" Vol.II, summarizes the role of the Medieval Catholic Church in destroying the pagan doctrine of the "Great Year" which implies an eternal universe. Unlike in the pagan Greek cosmos, all bodies, heavenly and terrestrial, were now on the same footing. This made eventually possible to think that the slow fall of the Moon in his orbit and the fall of an apple on earth could be governed by the same gravitational law.

Keywords: Modern science, quantitative description, natural facts, rational, scientific knowledge, Pierre Duhem, doctrine of the "Great Year", heavenly and terrestrial bodies on the same footing, gravitation.

Nowadays many educated people think, only because they have been culturally conditioned to think so, that Christianity –and very specially, the Catholic Church- are and have always been hostile to science. Nothing further from the historical truth.

Modern science is characterized by an impressive capability to describe in quantitative terms an enormous variety of natural facts. From elementary particles to galaxies, to the universe as a whole.

Through differential equations and quantitative mathematical solutions, capable of matching the most precise experimental data. In dynamics there is a subtle continuous development [1] from Buridan to Copernicus, from Copernicus to Newton, from Newton to Einstein. In electrodynamics there is also a subtle continuous and logical development from Coulomb to Volta, to Ampere, to Faraday to Maxwell. And in thermodynamics there is continuous development

Address correspondence to Julio A. Gonzalo: Escuela Politécnica Superior, Universidad San Pablo CEU, Madrid; Tel: 34-91-547-0815; UAN: 34-91-497-4767; Fax: 34-91-497-8579; E-mail: julio.gonzalo@uam.es

from Carnot, Joule and Mayer to Clausius, Kelvin and Planck. As physicists, for all of them, the working of the physical laws is coherent and consistent, and the working of man's intellect, sometimes very laboriously, is capable of discovering order and harmony in nature.

In our contemporary Western civilization, a civilization with roots in medieval Christendom, generation after generation of curious observers and a handful of geniuses have made a continuous progress and a series of momentous discoveries, following always a middle road between shortsighted positivism and an aprioristic idealism.

The intellectual adventure of building a scientific corpus in the Christian West was neither a random development nor a deterministic one, obviously.

If the world had not been made rationally, systematic scientific knowledge would be impossible. If something rationally valid here and now could be false at another place and another time, a substantial effort to investigate and study it would not be justified; physics as well as metaphysics presuppose the existence of an objective reality independent of the observer.

It does not require a scientist to see the truth that the world is rationally made. The astonishing edifice of contemporary science is proof that the laws of nature can be quantitatively and systematically described. They are, therefore pointers to the wisdom of the Creator of that nature and that laws which govern it. At the same time, science is a monumental proof that man has been endowed by the Creator with intelligence and freedom to investigate nature and nature's laws. The Christian West is the heir of Jewish biblical wisdom and of Greek and Roman early intellectual achievements. After the decline and fall of the Roman Empire, physical science took yet many delays and detours to be born finally in a Medieval Christian matrix. In this matrix, at long last, cosmic "eternal returns" and the pagan doctrine of the Great Year could be left behind.

At the turn of the nineteenth century and in the first years of the twentieth, Pierre Duhem, who had made already great contributions to theoretical physics in mechanics and thermodynamics (and had written a penetrating philosophical interpretation of physical laws) discovered, to his own surprise, the medieval origins of Newtonian physics at the University of Paris in the first half of the fourteenth century.

Figure 1: Pierre Duhem (1861-1916).

Duhem's extraordinary contribution has been somewhat reluctantly recognized in secularist academic circles but has been largely ignored in nominally Christian university campuses. In his book [2] "Scientist and Catholic: Pierre Duhem", Father Jaki gives a vivid portrayal of the dramatic life and work of Pierre Duhem. In the second half of the book he offers a representative sample of selections of Duhem's writings which illustrate well the unity accomplished in him of his science and his Catholic faith. R. P. Feynman [3], not a believer himself, said something which undoubtedly applies to Pierre Duhem: "many scientists do believe in both science and God - the God of revelation - in a perfectly consistent way". Historical perspective will keep competent physicists as far away from radical dogmatism as from radical skepticism.

In his "Le systeme du monde... Tome II. La Cosmologíe hellenique (1914), pp. 390, 407-408 (1914-1)" Duhem summarizes the role of the Medieval Catholic Church in destroying the pagan doctrine of the "Great Year" which, of course, implies an eternal universe, difficult to reconcile with the Big Bang.

> *"In the system which Maimonides sets forth we see, so to speak, the culmination of all the ideas whose development has been traced in this chapter.*
>
> *We find there, first of all, the affirmation of the principle that Aristotle had already formulated with such clarity: The various parts of the universe are interconnected by a rigorous determinism and this determinism subjects the entire world of generation and corruption to the rule of celestial circulations.*
>
> *We find there the corollary of that principle, namely, the definition of an astrological science which ties all changes accomplished here below to the motion of a specific planet.*
>
> *We see there the preponderant role which that astrology attributes to the Moon as a rule of water and humid matter. The Moon forces them to grow and decrease with her. The theory of tides clearly proves the reality of this lunar action and, through it, of all influence emanating from the celestial bodies.*
>
> *Finally, we hear stated that the very slow changes on earth are tied to the almost imperceptibly slow motion of the fixed stars whose revolution measures the Great Year.*
>
> *To that system all disciples of Greek philosophy –Peripathetics, Stoics, and Neoplatonist– have contributed. To that system Abu Masar offered the homage of the Arabs. The most illustrious rabbis, from Philo of Alexandria to Maimonides accepted that system.*

Christianity was needed to condemn that system as a monstrous superstition and to throw it overboard...

Hardly anxious to explore in detail the works of Greeks astronomers, the bishop of Hippo and with him, undoubtedly, the great majority of the Church Fathers, did not know how to separate, in a precise manner, the hypotheses of the astronomers from the astronomers' superstitions. The former were confusedly included in the disapprovals accorded to the latter...

Let us not therefore search in the writings of the Church Fathers for the traces of a meticulously and sophisticatedly treated science. We assuredly cannot find them there at all.

Let us not, however, neglect the little they said about physics and astronomy.

The first of their teachings on this topic are the first seeds from which the cosmology of the Christian Middle Ages would slowly and gradually develop.

Also, and above all, the Church Fathers hit, and did so in the name of the Christian Creed, the pagan philosophers on points which, today, we consider more metaphysical than physical but where actually lie the cornerstones of the physics of Antiquity: Such are the theory of eternal prime matter, the belief in the stars domination over sublunary things and in the periodic life of a cosmos subject to the rhythm of the Great Year. By destroying through these attacks the cosmologies of peripatetism, of Stoicism and of Neo-Platonism, the Fathers of the Church clearly prepared the way for modern science".

As pointed out above, one can trace [4] the continuous development of physics, or the science of massive bodies in motion, from Buridam to Copernicus, and from Copernicus to Newton.

Copernicus leaned about Buridan ideas in his critical commentaries to Aristotle's cosmological work, "On the Heavens", which he studied at the University of

Cracow, where the university library still has a dozen manuscript copies of Buridan's manuscript. Such copies can be found in many other significant European medieval universities. Buridan's ideas were further developed by his discipline, Nicole Oresme, who succeeded him in the Sorbonne.

The passage previously quoted of Buridan [5] makes clear its Christian theological matrix: belief in creation out of nothing and in time. This belief was held explicitly from early patristic times, and it was defined in 1215, at the Fourth Lateran Council. Both Aristotle and Ptolemy, as all other scholars of classical pagan times, were eternalists. The world had no beginning and no end, and, if it had large-scale changes they were only for the duration of a Great Year, or the duration of the full precession of the equinoxes, about 26000 years.

Today, three hundred years after Newton, inertial motion looks very natural. Why did that idea come so late? To answer this question it is necessary to take a look to Aristotle's ideas on motion. According to him the heavenly sphere moves because it is animated by its desire for the Prime Mover. But it would be a mistake to take Aristotle's Prime Mover for a Creator transcendent to the universe. As pointed out by Fr. Jaki, Aristotle spoke often as a pantheist for whom the universe was not only the Supreme Being, but the supreme living being.

The "animization" of the world is the essence of Plato's and Aristotle's systems, the two best followers of Socrates. It was very difficult if not impossible to formulate the laws of motion in nature, beginning with the law of inertia, in an "animated" world. So, according to the penetrating analysis of Fr. Jaki, the coming of modern physics (anticipated as shown above by Buridan, and then by Copernicus, Galileo and Kepler) was delayed until a Christian theological matrix could produce an intellectual climate according to which nature was created and was not the ultimate and supreme living being.

As a result, physical science witnessed only one viable birth: in Western medieval Christianity.

The medievals [6], and this may surprise many, made tremendous advances in producing technical devices. Until the advent of the steam engine, the Western world lived on technological innovations made during the medieval centuries. Among these innovations one should mention the <u>cam,</u> which allows the transformation of circular motion into linear motion and *vice versa* (therefore, the power provided by watermills could be used to drive mechanical saws, *etc.*) and the transformation of <u>accelerated</u> motion to motion at <u>constant velocity</u> (which made possible the construction of pendular clocks by the end of the thirteenth century, a vast improvement for time measurements).

For the medievals, the Christian belief in the Creator allowed a breakthrough in thinking about nature. Only a truly transcendental Creator could be powerful enough to create a nature with autonomous laws without his power being thereby limited.

Unlike in the pagan Greek cosmos, there could be no divine bodies in the Christian cosmos. All bodies, heavenly and terrestrial, were now on the same footing. This made eventually possible to think that the slow fall of the Moon in his orbit and the fall of an apple on earth could be governed by the same law of gravitation.

CONFLICT OF INTEREST

The author(s) confirm that this chapter content has no conflicts of interest.

ACKNOWLEDGEMENT

Declared none.

REFERENCES

[1]   Gonzalo.J.A, "The Intelligible Universe" (Singapore: World Scientific,2008).
[2]   Jaki S.L., "Scientist and Catholic: Pierre Duhem" (Front Royal, VA: Christendom Press, 1991).
[3]   "The Relation of Science and Religion", Engineering and Science 20 (June, 1956), p. 20.
[4]   Jaki.S.L, "Christ and Science" (Royal Oak, Michigan: Real View Books, 2000) p. 13.
[5]   Ibidem. p. 15.
[6]   Ibidem. p. 22.

CHAPTER 9

The Post-Renaissance Revolution: The New Science

Manuel M. Carreira S.J.

Universidad Pontificia de Comillas, Madrid

Abstract: A new approach to thinking about nature was developed in post-Renaissance Europe embracing an ever increasing body of theoretical and technical knowledge. It was accepted that there was and there is evolution both in "inert" and in living mater. This approach implied interactions ruled by "laws" not externally imposed. It applied to Astronomy, Geology, Physics, Chemistry, Biology and established the basis for the industrial revolution and changed the approach to Medicine, Economy and the transmission of Culture. At earlier periods the human experience of the world was interpreted in terms of mythological and religious models. Then in terms of scientific (formal) geometrical models and finally in models based upon scientific causality (first the mechanical model and then the dual model encompassing Relativity and Quantum Mechanics). The current success of astrophysics can only be expressed in the context of the "cosmological principle": the universe is homogeneous and isotropic. We can extrapolate our laboratory experiment under the same physical laws to infer the part and predict the future. But science has a proper subject and its own limits.

Keywords: Post-Renaissance Europe, a new approach to nature, science as the basis for the scientific revolution, mythological and religious models, scientific models, cosmological principle, the limits of science, a calendar of selected dates related to science.

During a time period centered between 1550 and 1700, a new way of thinking about material nature was developed in post-Renaissance Europe. This new approach, that embraced an ever increasing body of theoretical and technical knowledge, forms the framework of Modern Science. It is the outcome of a new "model", a viewpoint that made possible -through apparently insignificant advances- the attainment of the old dream of understanding the Universe by finding logical reasons for its structures and interactions both at the astronomical and at the microscopic levels, something that the more ambitious efforts to attain a single abstract explanation had failed to provide.

*Address correspondence to Manuel M. Carreira: Universidad Pontificia de Comillas, Madrid; Tel 34-91-540-6101; Fax 34-91-372-0218; E-mail: ecarreira@res.upco.es

Perhaps the deepest reason -not always consciously pursued- was that the entire material reality ceased to be considered as a single whole, made and structured once for all, accepting instead that there was and is evolution both in "inert" and in living matter. This way of thinking implies the need for interactions, activities that are ruled by "laws" that are simple statements of how things occur due to the nature of matter, not to any kind of external imposition This reasoning process is applied in Astronomy, Geology, Physics, Chemistry, Biology with consequences of enormous theoretical and practical impact that establish the basis for the industrial revolution and that change the approach to Medicine, to the Economy to the transmission of culture. Thus we find the new intellectual atmosphere as something we now accept as "obvious" but that we really must admire, mostly for its rapid development within human history.

In order to better understand this entire process it will be useful to remember how the way of thinking evolved within Western civilization, without forgetting that in other areas -the Orient especially- important steps were taken as well, mostly of a technical order We cannot give a complete historical account, but we shall underline moments and personal contributions that appear as critical through more than twenty centuries. Thus we can specify some models or viewpoints for which we have historical references.

MYTHOLOGICAL AND RELIGIOUS MODELS

In ancient civilizations for which we find sufficient archaeological evidence to infer their way of thinking regarding the material world (Babylon, Egypt, India, the peoples of central America among others) we uncover an early period when the human experience of the world is interpreted along polytheistic religious ideas We can't always clearly distinguish pantheism from polytheism and animism: in all of those there is a basic thought that accepts a kind of divine character for everything outside the ordinary experience of human life (even if an existence after death somehow changes also our own nature into something supernatural).

The material world -apparently quite inert- is really considered as endowed with some mysterious free and vital activity tied to it: there are spirits or gods of the air. The fire the rivers, the mountains, the trees, the sea, the storms. Even animals, especially those that are either dangerous or most useful, are in some way divine or linked to a divinity the snake the bull, the jaguar, the eagle. The divine dignity is especially assigned to those things that are inaccessible or impressive and mysterious: the Sun and the Moon, the planets volcanoes the Earth itself.

Figure 1: Isaac Newton (1642-1727).

Activities in nature are not considered as due to properties of a matter endowed with forces and fixed ways of acting: free superhuman beings determine everything that happens from the daily sunrise to the unexpected turns of human life Instead of Astronomy we find Astrology: an effort to interpret the plans of the gods for each individual, and to use magic to control future events that lack all sense of order and security. In many instances there is a background of rivalries among those entities that control the world in most mythologies the formation of the Earth and the existence of humanity are presented as the result of fights of the divine beings so that those who are overcome are the prime matter that is used for the several levels of material reality.

We must stress that at this stage nothing is said about the reason for the existence of mater (implicitly considered as primordial and eternal) and no indication is given of an evolution the Universe as a physical system. Rather, an initial chaotic stage is the first realty are deities that spring from that chaos are the source of order that makes the Earth fit for human life. Instead of a final state due to the changes imposed on matter by its activity according to innate laws, sometimes it is accepted that the world will go back to a chaotic condition, perhaps followed by a new re-structuring in an endless process due to a kind of irrational and blind "fate" that controls even the gods. It is possible to find this view in Indian mythology in Central America.

As a consequence of this way of thinking, science is impossible: there is no way to know nature in a way that -with theories and laws- will allow us *to understand and predict* its activity Instead of science we have alchemy, divining by omens, magical spells, hidden forces, kind of knowledge that is privy to sages "initiated" in some group and who possess special powers There is no objectivity or unanimity, but a variety of rites changing from one culture to another, from one city to the next one, from one sage or school to a rival one.

For the same reason, there is no special interest in physical measurements. It becomes necessary to have an elementary math for weights and lengths used in daily commercial activities for architecture, for establishing boundaries of agricultural fields and distances between cities, but there is no arithmetic or algebra as a subject of study. The symbols for numbers are ill suited for theoretical developments numbers appear mostly as records of royal taxes; of population census or commercial exchanges We have an example of this utilitarian attitude in the Bible, where it is said -as a practical statement- that the length of a circumference is 3 times its diameter.

In Israel, where its religion is based upon a unique transcendental monotheism, nothing appears that would imply either a superstitious and magical attitude or an acceptance of divine attributes for nature and its elements There can be no real god except Yahweh, thus avoiding any kind of possible rivalry or fight at the level of the divinity In the more modern books of the Bible the concept of "creation" is explicitly introduced but matter is presented as devoid of intrinsic properties that would determine its way of acting: it is the decree of the Lord that the Sun should come up every morning and that the seasons should follow each other in the proper order for the good of mankind.

This is certainly a very valuable theological "model" centered upon a provident and just God wise and all-powerful, who does not act by arbitrary whims or any kind of necessity, but always acts moved by a loving concern He has made everything in an orderly way, and has imposed laws to his creation but they are laws imposed *from outside* (for instance, to make the seas remain within their boundaries) There is no real basis to develop science, but only a chance to discover more or less by sheer luck some useful healing properties *in plants or* other things To study the heavens, except to determine a religious calendar and observance, was a danger it seemed to lean towards the pagan obsession with astrology In different ways the entire culture of the ancient Hebrew people is only the description of its relationship to Yahweh at the historical, social and personal level without developing their own art (architectural or decorative) or any science The "Wise Man" is the one who knows Law -the way Yahweh has established for Israel to remain faithful to God- who knows practical things and who is skilled in solving riddles or interpreting the saying of older sages. Nothing remains of the Wisdom of Solomon as his proper contribution to human culture: when the Queen of Saba visits him, she is impressed by his insightful answers to her multiple questions, but no systematic body of knowledge is mentioned.

SCIENTIFIC –FORMAL-GEOMETRICAL MODELS

The Greek cosmological model, from the first philosophers and geometers, appears in history as an effort to "save the phenomena", describing positions and motions that have no known cause, but *no effort is made to even look for one.* Geometry is the prototype of science, mostly because its inferences can be visually presented, while even a simple sum is practically impossible to do on paper if we use Roman numerals (Greek notation was similar). Perhaps it is hard for us to realize the importance of a proper symbolic notation when we try to do even the simplest arithmetic; a square root, all algebra and the following developments are unthinkable in the context of Roman and Greek ways of representing numbers.

Thus we find that concepts of order and relationships described by the geometry of regular figures and solids are applied to the Earth and to the movements of the heavenly bodies. Cosmology is considered adequate if, through combinations of geometric elements, it can explain the way the heavens look and foretell the motions of the planets, without giving reasons why things are the way we observe them.

By strictly geometrical arguments the Greek scientists before Aristotle could give a proof of the spherical shape of the Earth and, three centuries before Christ, Eratosthenes found its circumference and diameter. At about the same time Aristarchus measured the size and distance to the Moon, comparing it with the Earth's shadow during an eclipse and relating its diameter to its apparent size in the sky (half a degree). He even tried a correct method to find the distance to the Sun, but without being able to measure angles with the required precision Aristarchus also proposed that the position and apparent motions of planets in the sky against the background of stars could be described in simpler terms supposing the Sun as their center it was not necessary to wait until Copernicus.

But the same geometrical reasoning led to think that the Earth did not move: an orbit tens of millions of kilometers in diameter (at least 20 millions according to Aristarchus) seemed to require that during the year some stars would show a different position against the background of the more distant ones: this could not be observed, thus leading to the idea that the observer on Earth was not moving. This basic reason was behind Ptolemy's system, even if it required the introduction of cycles and epicycles as a way to calculate the position of the planets just at the minimal level of naked eye observations.

A clear indication of the purely formal value of the proposed astronomical constructions is found in the concept of an orbital motion about an imaginary center, *where no physical object is found.* The planet must move around an abstract center (of an epicycle), that also moves in a circle around the Earth.

Further developments ended by making the center of this main circle fall also in empty space what we observe is the result of combining circles upon circles in a way that resembles how in modern times a graphical system can produce straight lines, triangles and rectangles, by combining circular motions. We can also represent a complex wave of any sound by the sum of sinusoidal curves of the correct amplitude and frequency.

Therefore it was not considered important to decide if the Earth and planets move around the Sun or if the Earth is motionless: there was no *physical* reason why one or the other should take place. The "center" was not a source of any force, but just a geometric point where we placed a compass to draw orbits for which no reason was known, and it was not even sought. With a purely abstract Physics (truly just a set of philosophical presuppositions) the structure of the Cosmos was described in terms of "elements" endowed with properties and positions that give only the qualitative impression of order. Four elements -earth, water, air and fire- constitute the material world at the earthly level, the environment where we find imperfection, change and decay. A fifth element -"quintessence"- constitutes the heavenly bodies, in a realm where immutable perfection requires strictly circular movements at constant speed and spherical polished bodies, that eternally shine because that is *proper to their nature*, of a dignity above that of any matter below the Moon.

We find the idea of "natural place" as the basic reason advanced to explain the hierarchical ordering of things: higher dignity requires a higher place. The Earth, the heavy and most obviously inert and crass element, must be at the bottom. The heavier a body is, the more it seems to seek the lowest possible point, from all directions, not because the Earth attracts it, but because that is where its natural tendency requires it to go. Against a current actual misconception, the position of our planet is not one of honor, but just the opposite: it occupies the center as the lowest point to which all falling bodies must go if left to themselves.

The geometrical model was the framework for the development of Astronomy almost until Newton. The Alphonsine Tables of Alphonse X of Castile, the models of Tycho Brahe, Copernicus and Kepler, are varied methods of describing the data with abstract constructions, more or less complicated, but *all without physical reasons* to support them. Perhaps Kepler, suggesting something like a magnetic influence of the Sun upon the planets, begins to point to interactions or forces in the modern sense. But his geometric obsession is still evident when he tried to relate the distances of the five known planets to the Sun to the five regular solid bodies of Euclidean geometry. He had to finally accept that the orbits were not circles, but ellipses. But the "distorted circle" was an ugly fact that he never quite felt comfortable with. Geometry was an almost poetic way of describing a world that the Creator had made with order and measurement, according to a biblical expression. The Greeks could also assert "The Divinity is a geometer".

The historians of science and philosophy can describe in detail how cultural trends may be related to the development of the sciences that we now consider as part of our modern world. We will simply mention here that there was a widespread acceptance of the idea of "influences" of the heavenly bodies upon the Earth and human life, due to some kind of unspecified "pre-established harmony": the positions of the planets determined the growth of plants and animals and even the formation of metals within the Earth, and a comet could announce or cause pestilences and other calamities.

The entire Universe was conceived as a tissue of real relationships, not understood or quantified, that were due to the Creator (in the Christian way of thinking) in order to help human life. In an empirical way, science develops toward Chemistry instead of the esoteric ideas of alchemy, and towards a Physics that is, first of all, Mechanics. The planets are studied more and more as material bodies, without any astrological connotations, even if they are beautiful in the sky and they serve to establish a suitable calendar (the week reflects the seven bodies that are observed moving against the background of the "fixed stars"). Thus the way is laid to find a new model: the Universe as a perfect "machine", even if the meaning of this word will evolve with time, but always keeping the basic idea of quantifiable interactions rooted in the nature of matter itself.

In the meantime, the Indian symbols for numbers and the positional notation in a decimal system are introduced in Europe by the Arabs. Later on we find the zero, the decimal point, symbols for addition, subtraction, multiplication, square root, the equal sign: steps that render computation faster and clearer, and that will lead to a simple way of detecting quantitative relationships between numerical data obtained from experiments.

Models Based on Scientific Causality

a) Mechanical Model

A new way of thinking about the Universe slowly grows from philosophical and scientific ideas around the 17th century. Instead of considering as the main property of matter the concept of "proper or natural place" -that should determine the position and movements of the different elements, even in the sky- we see the progressive acceptance of a modern viewpoint: matter is the same at all levels, and its activities and structures are due to "forces" intrinsic to it. Their reach and way of acting can be experimentally determined and mathematically described, with equations that are more flexible and meaningful than the geometrical drawings of previous times.

Galileo's work on moving bodies stresses that the proper scientific method is the quantitative experiment, something rarely present in the study of "Natural Philosophy", even if we do find some important instances in the middle Ages and even in Aristotle. It is the beginning of Mechanics, and it leads to a radical new idea: a body has no predetermined tendency to be at rest or in uniform motion. Thus we get the concept of *inertia,* together with that of *friction,* both used to explain the continuation of a motion once started and also the obvious fact that finally the motion stops. The accelerated motion of falling bodies establishes a mathematical relationship between space and time (Galileo: in free fall, the distance is proportional to the square of the time) even if the lack of an accurate clock makes exact measurements almost impossible.

The moment has arrived when a description of the Universe in clearly physical terms becomes possible, thanks to Newton's genius. His three laws of Mechanics establish a basis for Physics as important for the new science as Euclid's works had been for Geometry. The world is understood as a machine and its driving force is Gravity: for the first time an interaction is proposed where a body acts upon another (something suspected in magnetic effects) *without contact.* This is so daring and surprising that Newton himself didn't quite affirm it: his prudent wording was that "everything happens *as if* bodies attract each other with a force that is proportional to their masses and inversely proportional to the square of their distance".

The laws of gravitation and mechanics give a perfect explanation for planetary motions and the orbits they follow: the solar system is a magnificent "clock" of amazing perfection. The fall of a body towards the center of the Earth does not obey any natural "tendency" to occupy a place dictated by its dignity, but rather follows from the same force that keeps planets and satellites in their orbits. Still, Newton has to face some problems from Physics and Philosophy that he cannot solve.

Gravitation seems to act instantaneously, and at a distance. Newton proposed the existence of absolute space as an eternal container for bodies, but without any physical property. Such space should be infinite (without limiting borders or center) to avoid a gravitational collapse of all masses to the center. Time should also be infinite, since a beginning of time implies a previous time as well. Since Newton did accept the fact of creation from the Bible, he ended by identifying space and time with divine attributes, something that is not a concern of Physics. And to keep the planetary system from falling into chaotic disorder -due to perturbations of all bodies interacting—he accepted that a divine intervention was needed from time to time to maintain stable orbits.

Even with those limitations, Newton's work changed forever the scientific panorama. Laplace suggested gravitational condensation of a spinning cloud to explain the formation of the planetary system. Perturbations in the orbit of Uranus were sufficient to infer the existence of Neptune, and it was found where the math of Adams and Le Verrier had predicted its position. From reports of previous sightings, Halley for the first time correctly predicted the return of a comet, now named after him. The mechanical description was the most perfect example of order and certainty, and it seemed that all science should end up by explaining everything in terms of its laws, a dream that was realized to a great extent when sound, heat and even light seemed reducible to vibratory motions, basically similar to the way matter acts at the macroscopic level.

Gravitational theory allows us to derive and give a reason for Kepler 's Laws. It appeared possible to attain the dream of reducing Physics to Newtonian mechanics, where the concept of "force" is central: all forces, gravitational or mechanical, will produce accelerations even if other effects -due to other forces- have to be accepted to explain changes in the states and properties of matter. This view reaches our time when matter itself is defined in terms of its four forces or interactions: gravitational, electromagnetic (long range interactions), strong and weak nuclear (extremely short range).

The science of optics becomes fundamental for the study of the information contained in light received from the stars. We do not see because our eyes send any kind of ray towards the object (a common opinion until 1025) but because "something" material is sent by the star towards the observer. Newton thought that it was a stream of particles, but that theory was improved by Huygens, speaking of "waves" that end up by being electromagnetic, not mechanical. The color of light will indicate the temperature of its source, and spectral analysis will be the key to find out the composition and motions of the stars.

Indications of evolution are also found in the areas of Geology and Biology, with cogent reasons to think in terms of enormous time spans for the age of rocks and for life forms that no longer exist. Fossils -previously considered as mere effects of a playful nature- are now seen as petrified remnants of real living, but now extinct, animals and plants. It seems that we should accept the ancient saying of Greek philosophers: *panta rei,* everything changes, a concept that is finally overpowering when -just about a century ago- we came to realize that the Universe itself had a sudden violent beginning from which it has evolved to its present state. But that insight required an unsuspected rethinking of many "certainties", a new viewpoint that led to a more drastic revolution than any previous one.

At the same time that theoretical Physics was expanding its reach, the industrial revolution extolled the machine, constantly improved and diversified, giving scientists the necessary means to get more and more precise data in all their experiments. With that higher precision applied to angular measurements with better telescopes, it was finally possible (in 1838) to prove stellar parallax and to determine stellar distances, opening to astronomical study the universe outside the solar system. In the 20th century the structure of the Milky Way could be established, with our position within it, and we can now speak of innumerable other galaxies and of the way the whole Universe appears from our viewpoint.

One can sense the optimistic feeling that was common among scientists of the 18th and 19th centuries, frequently accompanied by an air of satisfied superiority with respect to other fields of intellectual work that cannot rely upon experimentation and measurement. The hero of this new world is Newton, and it seemed that all that remained to do was to improve the accuracy of the data ("get one more decimal figure") and apply that methodology to all levels of nature, from atoms and molecules to the vastness of space.

b) A Dual Model: Physical and Geometric (Relativity) and Quantum Mechanical

The Newtonian approach, seeking measurable activities and a suitable mathematical formulation, continued to give important fruits that we can only mention in general terms. The discovery of magnetic and electrical forces, the synthesis of electromagnetism in Maxwell 's equations and their application to the nature of light, the negative outcome of the Michelson- Morley experiments (Cleveland. 1887) and the new understanding of the atom from natural radioactivity, are the most important steps towards a more complete description of physical reality. But the new model is framed during a time of unbelievable change of ideas, centered upon the Theory of Relativity at the astronomical level, and Quantum Mechanics in the infinitesimal world of atoms, in the first half of the 20th century.

In 1916 Relativistic Cosmology, developed from Einstein's genial insights, explained gravitational interactions by accepting geometrical distortions of empty space, reacting to the presence of masses and determining orbits as the paths that twisted space imposes upon moving planets. For Newton, space was just a container, empty, inert, eternal and infinite, but now it appears as an active component of the material world, with a changing geometry -that requires going beyond Euclid and common sense- that implies a fourth spatial dimension, so that the observed tri-dimensional reality can be distorted towards it Time is also included as a physical parameter in the description of the structure and activity of matter: processes by which we can detect its flow occur at a different rate under the influence of gravitational or mechanical accelerations. We have to consider the whole universe as a physical system with manifold influences, with a global geometry and a necessary evolution of its tri-dimensional volume either expanding or contracting.

Those who delve deeply into General Relativity and the Cosmology based upon it feel very frequently that its conceptual simplicity and radical newness makes them admire its beauty even more than its mathematical brilliancy. It is said that when asked what his reaction would have been if the experimental tests were contrary to his theory, Einstein replied that he would have been saddened because the Creator had wasted the opportunity to do something intellectually so beautiful. It is the opposite of the remark attributed to Alphonse X when presented with the complex theory of cycles and epicycles: "I think that, if I had been present at the creation of the world, I could have suggested something simpler". Due to its compelling logic, even against the tenets of our common sense, the relativistic cosmology and the geometric nature of gravity were almost universally accepted with enthusiasm by the best scientists.

Einstein's model suggested, for the first time within scientific circles, that the universe is *finite but unlimited*, and that it *evolved* from an initial state -at a time in the past that we might be able to determine- to a future condition imposed by the play of physical forces (work of Friedman and Lemaïtre). A viewpoint that appeared so unexpected and daring, that Einstein himself found it instinctively repelling and incompatible with his ideas or science, even if later he surrendered to experimental proofs.

The central idea of the new Cosmology might be found in the concept of multiple and universal interactions carried to its logical consequences: at all levels it describes activities of a unique reality that shows itself in multiple ways Explicitly, it involves the equivalence of mass and energy, so that "matter" includes particles and waves, the physical vacuum and even space and time.

While Relativity dealt with the totality of the Universe, Quantum Mechanics - to which Einstein also contributed, together with Max Planck, Bohr, De Broglie,

Heisenberg, Pauli, Dirac and Schrödinger, mentioning only the best known- presented us with an intimate description of matter impossible to reconcile with our intuitions The development of experimental physics, of the atom and its basic elements, leads to a synthesis of physical activity as due only to the four forces previously mentioned, that have different intensities, ranges and effects upon "elementary" particles that appear more and more as anything but simple and that cannot be described in terms of everyday language.

The nature of light -particles or waves? is stated without a clear answer: it is "something" that can manifest itself as a particle or a wave, depending upon the experiment we perform. The same answer must be given regarding the more "material" components of ordinary bodies atoms and molecules can behave as waves We cannot distinguish the effect of gravitation from that of a mechanical acceleration in a spaceship, and we certainly are at a loss when told that energy (always considered as something "immaterial", an *accident* and not a substance) can be changed into particles, and also the opposite.

At present we still lack a unifying synthesis between the viewpoint of continuous fields of Relativity and the discontinuity of energy -and even the space-time substrate- in Quantum Mechanics. The quest for unity reaches what one could consider a desperate level when proposing multiple Universes (clearly defined as undetectable and thus outside scientific methodology): we are presented with a new "mythology" of hypothetical constructions where hints from particle physics are expected to support the new cosmological hypotheses, that, in turn, must provide a reason to expect the existence and properties of those particles.

But -after many years- there are no data to lead us to accept those ideas and experimental verification appears impossible even in principle: we seem to go back to the way of thinking of ancient Greece when epicycles and sidereal substances provided a merely formal explanation of things that were not understood We now have, instead, multiple dimensions and mysterious types of matter or energy.

We thus arrive to the present, when we see no clear path towards a cosmology that will truly explain the world, at all levels, from a single viewpoint. We are forced to work with a methodology -based upon practical constraints and logically coherent- but without strict proof It is expressed in the so-called "Cosmological Principle": the Universe is homogeneous and isotropic There are no peculiar directions or places when we examine sufficiently large volumes. Therefore every observer anywhere would see the same structures in those surroundings and would attain a similar description of the world.

This confidence arises from the mechanical model carried to its limits, and it permits us to obtain far reaching consequences about the structure and evolution of the Universe It rests upon many observations and experiments that are understood within a framework of general statements: large volumes of the universe in opposite directions show the same variety and abundance of objects and structures (homogeneity); the spectral analysts of light from stars and galaxies indicate the same elements ruled by the same forces that determine wavelengths and energies proper to each element (universality and constancy of laws and material properties) Matter is the same at all places and we can extrapolate our laboratory measurements to the same elements everywhere under the same physical conditions Physical laws are universal and unchanging They can be applied to infer the past and to predict the future at any place and time, if we know the initial conditions and the laws of each system. This will be true, of course, within the margin of error of our measurements.

The current success of astrophysics can only be explained in the context of this cosmological principle as an objective description of the observable Universe.

THE SUBJECT AND LIMITS OF SCIENCE

In 1948 the great telescope of Mount Palomar was officially dedicated, and for many years remained as the largest and most productive in the world It was then said that "if the Universe that modern Science shows is truly admirable more admirable still is the power of the human mind that has discovered and analyzed so many marvels' Scientific work is not finished in any field each new experimental datum or observation gives rise to new questions It has been said also that "Science is born by answering questions and it develops by questioning answers".

Science, by its methodology, can only speak of whatever can be measured and that can be checked in an experiment that anybody having the necessary means can perform But prestigious scientists are aware of the limitations *of* that methodology that leaves outside the scientific treatment everything we call *the Humanities* and that embraces the most valuable aspects of human activity: the Arts. Ethics (human dignity rights and duties). Literature family and social relationships. It cannot deal even with the basic *desire to know* that produces Science with its never satisfied quest to find Truth Beauty and Goodness things that cannot be measured or introduced into a mathematical equation.

Abstract thought and free will cannot be described as properties *of* matter or explanation by the play of the four forces that define it Even when dealing with the material Universe the most basic question (according to J.A. Wheeler) is "Why is there something instead of nothing?" The next one, he says, is "What relationship exists between human existence and the properties of the Universe at the first moment?" And he goes on to admit that if we have no answer we should confess that we don't understand anything completely. But the final "why" and "what for" cannot be expressed mathematically or as the result of a measurement.

These limitations of the scientific method are unavoidable and they cannot be attributed to it as temporary limitations. On the other hand, they should not be considered a reason to undervalue or despise the sciences. Every type of human knowledge is partial, even if true, and thinkers from all times and cultures can contribute to the common treasure of human wisdom: there is no danger that we might exhaust the field of what can be known and understood. From S. Augustine to Newton and Einstein, the clearest minds have marveled at their own discoveries, while confessing the immensity of the boundless ocean they sensed lay behind their achievements. Only those who know very little can think that they know it all.

There is a story -for whose accuracy I can't vouch- regarding a letter from the man in charge of the Patent Office in the U.S towards the end of the 19th century, addressed to the Government that -as usual- found itself short of funds. The letter proposed closing the Office, "because everything had already been invented". And a well known scientist, towards 1925, tried to discourage a student who was planning to study university Physics: he should get into another field, "because there was physics left only for four or five years". There is still no danger of that.

"Nature, and Nature's laws lay hid in night. God said, *Let* Newton *be!*.and all was light".

Alexander Pope. English poet, eighteenth century.

"I do not know what I may appear to the world, but to myself I seem to have been like a boy playing on the sea-shore, and diverting myself in - now and then- finding a smoother pebble or a prettier shell than ordinary, whilst the great ocean of truth lay all undiscovered before me".

Sir Isaac Newton, English mathematician and physicist, c. 1700.

"One thing I have learned in a long life: that all our science, measured against reality, is primitive and childlike -and yet it is the most precious thing we have".

"You imagine that I look back on my life's work with calm satisfaction. But from nearby it looks quite different.There is not a single concept of which I am convinced that it will stand firm, and I am uncertain whether I am in general on the right track. I don't want to be right.I only want to know whether I am right".

Albert Einstein.

APPENDIX: SEVEN CENTURIES

A CALENDAR OF SELECTED DATES RELATED TO SCIENCE [1]

1054   - Chinese observers describe a Supernova in Taurus (now visible as the Crab nebula).

1500   - Leonardo da Vinci performs anatomical dissections.

1504   - Henlein makes the first pocket watch.

1514   - First version of Copernicus' model - Vesalius publishes "Anatomia Humana".

1540   - Rheticus publishes a summary of Copernicus' theory.

1543   - Full printing of Copernicus' work, after his death.

1550   - Geronimus Cardano proposes an evolutionary theory.

1572   - Tycho Brahe observes and describes a Supernova.

1574   - Tycho Brahe opens the first permanent observatory in the island of Hven.

1577   - Tycho Brahe describes a comet, placing it three times farther than the Moon.

1578   - Mexico University opens a Medical Department.

1582   - Gregorian Calendar introduced (based on work by Fr. Clavius, S.J.).

1586   - Stevinus shows that objects of different weights fall with the same speed.

1590   - Galileo publishes his mechanical experiments in "De Motu".

Janssen builds a compound microscope.

1600   - Gilbert describes the Earth as a magnet controlling the compass.

1604   - Kepler observes a Supernova, described in a book in 1606

        Galileo: In free fall, distance is proportional to the square of the time.

1605   - Francis Bacon: observation is the way to advance in science.

1608   - The terrestrial telescope is in public use.

1609   - Galileo uses the telescope in astronomy.

        Kepler publishes "Astronomia Nova" with two laws of planetary motion.

1611   - Sunspots observed by Galileo, Harriot, Fabritius and Scheiner S.J.

1617   - Briggs introduces logarithms of base 10.

1619   - Kepler publishes the third law of planetary motion in "Harmonice Mundi".

1620   - "Novum Organum" by Francis Bacon: the coasts of America and Africa fit 1624.

1624   - Pierre Gassendi measures the speed of sound in air.

1633   - Trial of Galileo in Rome.

1637   - Descartes introduces Analytic Geometry.

1645   - Von Guericke invents the air pump to produce a vacuum.

1656   - Huygens discovers that Saturn has rings.

1657   - Robert Hooke demonstrates that all bodies fall equally fast in a vacuum.

1665   - Newton develops his ideas on calculus, gravitation and optics.

1668   - Newton makes the first practical reflecting telescope (Newtonian).

1670   - For the first time clocks are made with minute hands.

1675   - Romer estimates the speed of light with a 25% error.

1678   - Huygens explains light as a wave (published in 1690).

1687   - Newton's Philosophiae Naturalis Principia Mathematics is published.

1728   - James Bradley discovers aberration of starlight, proving Earth's motion.

1744   - Lomonosov correctly interprets heat as a form of motion.

1758   - As predicted by Edmund Halley in 1705, Halley's comet returns.

Mathematical Symbols [2]:

876   :   The zero is used in India.

1202   :   Fibonacci introduces in Europe the Arabic (Indian) numerals.

1492   :   Francesco Pellos introduces the decimal point.

1514   :   First use of + and – signs.

1525   :   Christoff Rudolph introduces the square root symbol.

1557   :   Robert Recorde uses = sign

1631   :   William Oughtred introduces x sign for multiplication. Thomas Harriot uses raised dot and > and <.

1659   :   Symbol + for division introduced by Johann H.

CONFLICT OF INTEREST

The author(s) confirm that this chapter content has no conflicts of interest.

ACKNOWLEDGEMENT

Declared none.

REFERENCES

[1]   Ochoa.G, Corey M. "The Timeline Book of Science" (The Stonesong Press Inc.-N.Y. 1995).

[2]   Carey.J, ed. "Eyewitness to Science" (Harvard University Press: Cambridge, Mass. 1995).

CHAPTER 10

The True Pioneers of Modern Physics

Julio A. Gonzalo*

Escuela Politécnica Superior, Universidad San Pablo CEU, Madrid

Abstract: Lord Rayleigh comments on crude views about nature at the 54th Meeting of the British Association for the Advancement of Science. Many first rate scientists, true pioneers of physics, chemistry, astronomy *etc.* have opposed materialism and philosophical relativism. Both religion and science can claim objectivity if there is a world independent of human thought. Planck and Einstein affirm it unmistakably. Historically there have been some outstanding scientists leaning to agnosticism or materialism, like Mach, Poincaré and Bohr, but they are rather exceptions to the general rule.

Keywords: First rate pioneers of physics, first rate pioneers of chemistry, first rate pioneers of astronomy, objectivity of the world as viewed by Planck and Einstein, scientists leaning to agnosticism and materialism.

More than one hundred years ago [1], in a public speech before the 54th Meeting of the British Association for the Advancement of Science, Lord Rayleigh, one of the most distinguished British physicists of the time, said:

> *"Many excellent people are afraid of science as tending to* materialism *(emphasis added). That such apprehension should exist is not surprising, for unfortunately there are writers, speaking in the name of science, who have set themselves to foster it. It is true that amongst scientific men, as in other classes, crude views are to be met with as to the deeper things of nature; but that the lifelong beliefs of Newton, of Faraday, and of Maxwell, are in consistent with the scientific habit of mind, is surely a proposition which I need not pause to refute."*

In Germany, in France, all over Europe, and in the United States, on the other side of the Atlantic, at the time, many writers, announcing themselves as champions of

*Address correspondence to Julio A. Gonzalo:** Escuela Politécnica Superior, Universidad San Pablo CEU, Madrid; Tel: 34-91-547-0815; UAN: 34-91-497-4767; Fax: 34-91-497-8579; E-mail: julio.gonzalo@uam.es

science, proclaimed in its name and on its authority, as noted by Fr. Karl Kneller, SJ [2], the imminent defeat of religion, more explicitly the Christian religion.

So the wave of books leading to agnosticism or atheism by prominent scientists (like Monod and Watson), or very successful authors of popular scientific books (like Asimov and Sagan) is <u>nothing new</u>.

<u>But, Is the generalized impression that science supports atheism well founded?</u> To show that it is not, something that Lord Rayleigh considered <u>unnecessary</u> one hundred years ago, we may document the fact, as done by Fr. Kneller, that many first rate scientists, true <u>pioneers</u> of physics, chemistry, astronomy, etc, have opposed materialism, as well as relativism, from a theist perspective, or from practicing Christian convictions. To this end we may implement the tactic of Diogenes, who, when confronted with the Sophists' claim that it was impossible to actually move from one point in space to another, did walk in silence from one side of the room to the other, refuting so the Sophists' claim by the method of "solvitur ambulando".

The same could be easily done about the claim that to be a scientist means to be a materialist, or an atheist, or a pantheist. To that end one merely needs to list a representative sample of believing scientists who were, at the same time, truly great scientists.

As noted by Stanley Jaki [3], both <u>religion</u> and <u>science</u> can claim objectivity only if there is a world existing independently of human thought. This <u>objectivity</u> was not more unmistakably affirmed in our times than by the two greatest physicists of the twentieth century, Plank, the pioneer of quantum physics, and Einstein, the author of the theory of relativity, special and general.

Sir William Bragg, who was awarded the Physics Nobel Prize in 1915, said about science and religion: "They are [opposed] in the sense that the thumb and fingers of my hand are opposed to one another. It is an opposition by means of which anything can be grasped".

The cultural impact of Marxism (a most unscientific materialism), together with that of crude materialistic evolutionism, since the second half of the nineteenth century, as well as that of a certain pantheistic Neo Platonism among theoretical physicists one century later, has contributed to give the impression that the scientific enterprise, three centuries after Newton, is somehow connected with atheism or pantheism.

Figure 11: Cover of "Dios y los científicos" by Julio A. Gonzalo.

Let us put together a sample of representative quotations by outstanding pioneers of all branches of the physical sciences [4] which demonstrates that this is false:

*In Classical Mechanics:

Isaac Newton (1642-1727): "This most beautiful system of the sun, planets and comets could only proceed from the counsel and dominion of an intelligent and powerful Being".

*In Electricity:

Alexander Volta (1745-1827): "I do not understand how anyone can doubt the sincerity and constancy of my attachment to the religion which I profess, the Roman, Catholic and Apostolic religion, in which I was born and brought up. I constantly give thanks to God Who has infused into me this belief in which I desire to live and die with the firm hope of eternal life".

André Marie Ampère (1775 -1836): "We can see only the works of the Creator but through them we rise to the knowledge of the Creator himself. so God is in some sort hidden by His works, and yet it is through them that we discern Him and catch a hint of the Divine attributes".

Michael Faraday (1791 -1867): "There is no philosophy in my religion. I am of a very small and despised sect of Christians. But, though the natural works of God can never by any possibility come in contradiction with the higher things that belong to our future existence. I do not think it at all necessary to tie the study of natural sciences and religion together. that which is religious and that which is philosophical have ever (been for me) two distinct things" (Letters, October 24th,1844)

James Clerck Maxwell (1831 - 1879): "I have looked into most philosophical systems and I have seen that none will work without a God" / "Old chap! I have read up many queer religions: there is nothing like the old think after all".

*In Thermodynamics:

Julius Robert Mayer (1814 - 1878): He maintained ". that in the living brain physical changes designated by the name of molecular activity are continually taking place, and that the spiritual functions of the individuals are connected in the most intimate manner with this cerebral processes. But it is a gross blunder to identify these concurrent activities. An example will demonstrate it unequivocally. Telegraphic communication cannot, as everyone knows, be established without a simultaneous chemical process. But the message delivered by the wire, the contents of the telegram, can by no means be regarded as a function of this electrochemical process. This holds with still grater force of the relation of thought to the brain. The brain is not the soul but only the instrument of the soul. not an object of investigation for Physics and Anatomy. Were it not for this inalterable harmony, pre-established by God, between subject and object, all our thinking would be necessarily without fruit. Logic is the Statics, Grammar the Mechanics and Language the Dynamics of thought".

James Prescott Joule (1818 - 1889): "I shall loose no time in repeating and extending these experiments, being satisfied that the grand agents of nature are, by the Creators Fiat, indestructible, and that whatever mechanical force is expended, an exact equivalent of heat is always obtained".

William Thomson (Lord Kelvin) (1824 - 1907): "It is impossible to understand either the beginning or the continuance of life without an overruling creative power": "I need scarcely to say that the beginning and maintenance of life of Earth is absolutely and infinitely beyond the range of all sound speculation in dynamical science. The only contribution of dynamics to theoretical biology is absolute negation of automatic commencement of life".

*In Optics:

Josep Fraunhoffer (1787 - 1826): - "Fraunhoffer was a man of disciplined and benevolent temper, occasionally clouded, it is true, by outbursts of his natural irritability. He was a loyal adherent of his religion (Catholicism), so thorough in his obedience that even those invited to this house were obliged to observe the prescribed fasts and abstinences, a remarkable contrast, surely, to the license of his days" (Historich-politische Blätter XI, Munchen 1843, 485).

Augustin Fresnel (1788 - 1827): - "He saw coming his end with the religious sentiments of a man who, having been carried further than his fellow men into the secret of the marvels of nature, was deeply penetrated by the infinite power and goodness of its author" (Oevres Completes III, 525 - 526).

Armand Hippolyte Louis Fizeau (1819 - 1896):- "Fizeau was all his life a loyal, open and practical Christian. It was for this reason his name was struck of the list of those presented for the Cross of the Legion of Honor at the Centenary celebration of the Academy. Cornu's was the only voice in protest against the "odious and revolting intrigue" (Private communication to K. A. Kneller, op. cit. pp. 158 - 59).

*In Chemistry:

Robert Boyle (1627 - 1691): Author of a book with this expressive title - "The wisdom of God manifested in the works of his Creation".

John Dalton (1766 - 1844): - "In Dalton the character of the man equaled the superiority of his lights: he was a model of virtue without ostentation, of religion without fanatism". When he died, more than 40.000 people came to pay the last tribute to his mortal remains. His disgust with the atheistic systems which begun to be popular in his time was evidenced in his great chemical treatise: "New System of Chemical Philosophy". (See K. A. Kneller op. cit. p 180).

Jöns Jacob Berzelius (1779 - 1848): "An incomprehensible force, foreign to those of dead matter, has introduced the principle (of life) into the inorganic world. This has been effected not by chance, but by the striking variety and supreme vision of a plan designed to produce definite results, and to maintain an unbroken succession of transient individuals which are born one from another, and which at their death bequeth their decomposed constituents to the formation of new organisms. Every process of organic nature proclaims a wise purpose, and bears the stamp of a guiding mind; and man, comparing the calculations he makes to attain certain ends with those he finds in organic nature has been lead to regard his faculty of thought and calculation as reflection of the Being to whom he owes his existence. And yet it has happened more than once that a philosophy, proud all the while of its own profundity, has maintained that all this was the works of chance, and that such organisms only held their ground as had accidentally acquired the power of self-preservation and reproduction. But the advocates of such systems do not perceive that the element in nature that they call "chance" is a thing physically impossible. Everything which exists springs from a cause, an operative force, this latter tending (like desire) to break into activity and secure for itself satisfaction. a process which in no degree corresponds to our idea of 'chance'. It will be more honorable for man to admire the wisdom which he cannot rival, than to puff himself up with philosophical arrogance and attempt with his paltry reasoning to penetrate mysteries which will probably remain for ever beyond the scope of human reason".

*In Mathematics:

Karl Friedrich Gauss (1777 - 1855): "There is in this world a joy of the intellect which finds satisfaction in science, and a joy of the heart which manifests itself in the aid men give one another against the troubles and trials of life. But for the Supreme Being to have created existences and stationed them in various spheres in order to taste these joys for some 80 or 90 years - that were surely a miserable plan.". "We are thus impelled to the conclusion to which many things point, although they do not amount to a coercive scientific proof, that besides this material world there exists another purely spiritual order of things, with activities as various as the present, and that this world of spirit we shall one day inherit".

"The indestructible idea of personal survival after death", says a biographer, "the steadfast belief which he had in a Supreme Ruler, a just, eternal, omniscient, omnipotent God, formed the foundation of his religious life, and in unison with his matchless scientific achievements formed a perfect harmony".

Augustin Louis Cauchy (1789 - 1857): "I am a Christian, that is to say, I believe in the divinity of Jesus Christ as did Tycho Brahe, Copernicus, Descartes, Newton, Fermat, Lebnitz, Pascal, Grimaldi, Euler, Guldin, Boscovich, Gerdil; as did all the great astronomers, physicists and geometricians of past ages: nay, more, I am like the greater part of these a Catholic: and were I asked for the reasons of my faith I would willingly give them. I would show that my convictions have their source not in mere prejudice but in reason and resolute inquiry. I am a sincere Catholic as were Corneille, Racine, La Bruyère. Bossuet, Bourdaloue, Fénelon, as were and still are so many of the most distinguished men of our time, so many o those who have done most for the honor of our science, philosophy and literature, and have conferred such brilliant luster on our Academies. I share the deep conviction openly manifested in words, deeds and writings by so many savants of the first rank, by a Ruffini, a Haüy, a Laënnec, an Ampere, a Pelletier, a Freycinet, a Coriolis and I avoid naming any of those living for fair of pining their modesty. I may at least be allowed to say that I love to recognize all the noble generosity of the Christian faith in my illustrious friends the creator of Crystallography (Haüy), the introducers of quinine and the stethoscope (Pelletiesr and Laënnec), the famous voyager on board of the 'Urania', and the immortal founders of the theory of Dynamic Electricity (Frencinet and Ampère)".

Bernand Riemann (1826 - 1866): - He died of tuberculosis not yet forty years old. In the biographical sketch prefixed to his collected work, the following is registered: "His wife was saying to him the Our Father; he was no longer able to speak, but at the words "Forgive our trespasses", he raised his eyes aloft, and after two or three gaps, his noble heart ceased to beat. The pious habits which he had learned in this childhood remained with him all his life, and he served God loyally, if not always after the orthodox forms. Daily self-examination in the presence of God he regarded. as one of the elements of religion".

*In Astronomy:

William Herschel (1738 - 1822): ". by metaphysics (those of us who love wisdom) are enabled to prove the existence of a first cause, the infinite author of all dependent beings".

Heinrich Wilhem Matthaüs Olbers (1758 - 1840): "I am grateful for the easy circumstances with which Providence has blessed me. I (have) tasted and enjoyed all good that this earthy life has to offer and (I am) now able to take my departure without reluctance" (Letter to his friend Frederick W. Bessel, in his old age).

Frederick W. Bessel (1784 - 1846): "Let us enjoy what God, Whose goodness is so infinitely beyond that of man, has given us [.] God knows how hard is for me to be so near to you and yet not to be able to visit you".

Urbain Jean Joeph Leverrier (1811 - 1877) - "It was given to him, said Dumas (at his funeral) to write the last word of the last page of his immortal work in the last hour of his life, murmuring as he concluded: 'Nunc dimittis servum tuum, Domine'; ". the study of heaven and the scientific faith in him did nothing else but consolidate the living faith of a Christian".

A list of outstanding materialists or atheist scientists, since the times of Newton to this day, could no match, for any impartial observer, the above list of convinced theist, mostly Christian, pioneers of the physical sciences given above.

Regarding the twentieth century [5], it is not difficult to make a list of distinguished Nobel Prize winners, Protestant, Catholic, Jew, who were neither materialists not atheists.

*In Physics:

Planck (the father of Quantum Physics), Marconi (pioneer of radio waves telecommunication), de Broglie (first proponent of the wave-particle duality), Compton (discoverer of the Compton effect, which demonstrates the dual particle-wave character of light), Hess (pioneer of the cosmic rays), Penzias (co-discoverer of the cosmic background microwave radiation), Townes (discoverer of the "maser"), Mainman (discoverer of the "laser"), Brockhouse (pioneer of neutron spectroscopy);

*In Chemistry:

Rutherford (father of nuclear physics, buried in Westminster Abbey, near the tomb of Newton), G. P. Thomson (discoverer of the electron), Debye (with major theoretical contributions to dipole moment theory, to the low temperature specific heat of solids, to electrolytic solutions, to quantum chemistry, etc, etc), Seaborg (discoverer of ten transuranic elements), Hinselwood (author of important theoretical contributions to the kinetics of chemical processes) Perutz (discoverer of the structure of hemoglobin and globular proteins).

Other outstanding twentieth century physicists who were not awarded the Nobel Prize are Lemaitre (father of the Big Bang concept) Sandage (who began his career as assistant of Edwin Hubble, at he Palomar Observatory), von Braun (the most preeminent rocket engineer of the 20th century who developed the Saturn V booster rocket, making possible to land the first men in the Moon in 1969).

Of course there have been outstanding pioneers of science leanings towards agnosticism or materialism, like <u>Mach</u>, <u>Poincare</u>, <u>Bohr.</u>

But they are rather <u>exceptions</u> which confirm the general rule.

CONFLICT OF INTEREST

The author(s) confirm that this chapter content has no conflicts of interest.

ACKNOWLEDGEMENT

Declared none.

REFERENCES

[1]   Kneller K.A.,"Christianity and the Leaders of modern Science" (Fraser, Michigan: Real View books, 1995) p.I, note1.
[2]   Ibidem: Introduction, p. vii
[3]   Ibidem, p. xxi.
[4]   Ibidem, Retrospect, pp 387-400.
[5]   See f.i. Julio A. Gonzalo, "The Intelligible universe" (Singapore: World Scientific, 2008).

CHAPTER 11

A Finite, Open and Contingent Universe

Julio A. Gonzalo[*]

Escuela Politécnica Superior, Universidad San Pablo CEU, Madrid

Abstract: According to today's Cosmology the universe has a finite mass, a finite (but growing) "age", and a finite (but growing) space-time extension. The product of H_o (Hubble's parameter) by t_o (the present "age" of the universe) is at the present epoch $H_{ot}t_o = 0.942 \pm 0.065$, therefore more than 2/3 implying an open universe (k<0). A finite, open universe is contingent (it could have been otherwise) and therefore created. Quotes of Planck and Einstein on the subject are given.

Keywords: Cosmic finiteness, cosmic mass, cosmic "age", cosmic extension, observational evidence for an open (k<0) universe, Plank's and Einstein's views on the subject.

The universe is <u>finite</u> (see f.i. Julio A. Gonzalo, "Inflationary Cosmology Revisited" (World Scientific: Singapur, 2005) [1]": As we have seen, it has a finite <u>mass</u>

$$M = 1.54 \times 10^{54} \, gr \,,$$

a finite (but growing) "<u>age</u>" since the Big Bang,

$$t_o = 13.7 \times 10^9 \, yrs \,,$$

and a finite (but growing) <u>radius</u>, since that event,

$$R_o = 9.96 \times 10^{27} \, cm \,,$$

The finite mass $M = 1.54 \times 10^{54} gr$. is conserved, of course through the cosmic expansion, while the actual time and the actual radius are finite, but increasing

---

*Address correspondence to Julio A. Gonzalo: Escuela Politécnica Superior, Universidad San Pablo CEU, Madrid; Tel: 34-91-547-0815; UAN: 34-91-497-4767; Fax: 34-91-497-8579; E-mail: julio.gonzalo@uam.es

and, in and <u>open</u> universe, unbounded. We may note [2] that the characteristic radius R_+, for a universe with a total matter mass given by $M = 1.54 \times 10^{54} gr$, with $y_+ = \sinh^{-1}(1)$, is given by

$$t(y_+) = \frac{R_+}{|k|^{1/2} c} \{\sinh y_+ \cosh y_+ - y_+\} = 0.365 \times 10^9 \, yrs \,,$$

$$R(y_+) = R_+ \sinh^2 y_+ = 4.58 \times 10^{26} \, cm \,.$$

A <u>Compton</u> radius

$$r_c = \frac{\hbar}{mc} \,,$$

and a <u>Schwarzschild</u> radius

$$r_s = \frac{Gm}{c^2} \,,$$

can be associated to the universe as well as to various finite objects in it, going from finite galaxies to finite elementary particles.

The following Table gives <u>Compton</u> radius and <u>Schwarzschild</u> radius for various massive objects in the universe.

Compton and Schwarzschild radii for massive objects

Object	m(g)	r_c(cm)	r_s(cm)	r_c/ r_s
Universe	$1.54 \cdot 10^{54}$	$2.26 \cdot 10^{-92}$	$1.14 \cdot 10^{26}$	$1.98 \cdot 10^{-118}$
Galaxy	$\sim 1 \cdot 10^{43}$	$3.5 \cdot 10^{-81}$	$7.41 \cdot 10^{-14}$	$0.47 \cdot 10^{-95}$
Star	$\sim 1 \cdot 10^{32}$	$3.5 \cdot 10^{-70}$	$7.41 \cdot 10^{3}$	$0.47 \cdot 10^{-73}$
Earth	$5.95 \cdot 10^{24}$	$5.85 \cdot 10^{-63}$	$4.43 \cdot 10^{-4}$	$1.32 \cdot 10^{-67}$
Planck m.p.	$2.17 \cdot 10^{-5}$	$1.61 \cdot 10^{-33}$	$1.61 \cdot 10^{-33}$	$\underline{1}$
Baryon	$1.67 \cdot 10^{-24}$	$2.09 \cdot 10^{-14}$	$1.23 \cdot 10^{-52}$	$1.23 \cdot 10^{38}$
Electron	$9.10 \cdot 10^{-28}$	$3.84 \cdot 10^{-11}$	$6.74 \cdot 10^{-56}$	$0.56 \cdot 10^{45}$

If the Friedmann -Lemaitre solutions of Einstein's cosmological equations describe correctly cosmic evolution, (and they do describe well the thermal history of the universe from the Big Bang to present) the dimensionless product of the Hubble constant time the present "age" of the universe is given by

$$H_o t_o = 0.942 \pm 0.065,$$

which is incompatible either with a "flat" universe (k = 0), which requires

$$H_o t_o = 2/3,$$

or with a "closed" universe (k > 0), which requires

$$H_o t_o < 2/3.$$

The universe is therefore "<u>open</u>", finite and unbounded.

Modern physics tells us that we live in an evolving, finite and "open" universe.

Why is the Universe What it is According to Modern Cosmology, and No Something Else?

No purely physical theory and no purely physical experiment can give a concrete answer.

In other words the universe is "<u>contingent</u>", it is not <u>necessarily</u> what it is, but it is <u>really</u> what it is, and not anything else. The term "<u>contingent</u>" implies a <u>physical</u> reality which cannot be measured directly, but also a <u>metaphysical</u> reality, more intangible but no less real, which can be intellectually recognized.

And a "Contingent" Universe is a "Created" Universe

By the middle of the 19th century, both the Hegelian <u>left</u> (Marx and Engels) and the Hegelian <u>right</u> (specially the Neo-Kantians) had for a fundamental tenet that the universe (material or not) was <u>infinite</u>. Only a few first rate scientists dared to disagree. Among them <u>Gauss,</u> the prince of mathematicians, who noted that <u>Kant</u>'s "dicta" on categories were sheer triviality, probably keeping in mind non-Euclidean geometries. The finiteness of matter in endless space was implicit also in the work of <u>Riemann </u>and <u>Zöllner</u>.

Kant's <u>claim</u> that the universe was a bastard product of the metaphysical cravings of the human intellect (put forward to discredit the classic cosmological argument to prove God's existence) was flatly discredited with words and deeds by Planck and Einstein, the two greatest physicists of the 20th century.

Figure 1: Albert Einstein (1879-1955).

Planck, after liberating himself of Mach's tutelage, when he affirmed unambiguously (see f.i. S. L. Jaki, "The road of science and the ways to God" (Chicago U. Press: Chicago, 1978)) his full confidence in the reality of a <u>causally connected universe</u>.

<u>Einstein</u>, after finally emancipating himself of Mach's influence, when he produced the first contradiction-free treatment of the totality of all gravitational interacting things which explicitly required a <u>universe</u> <u>with a finite mass</u>.

When Planck's son, Erwin, was executed for plotting against Hitler at the end of World War Two, everything seemed to have fallen in ruins around him –home, country, science-. He wrote to a friend (see f.i. S. L. Jaki, Ibidem):

"What helps me is that I consider it a favour of heaven that since childhood
a faith is planted deep in my innermost being, a faith in the Almighty and

the All-good not to be shattered by anything. Of course his ways are not our ways, but trust in him helps us through the darkest trials"

In 1952, few years before his death, Einstein's wrote to his friend Maurice Solovine:

". I think of the comprehensibility of the world… as a <u>miracle</u> (emphasis added) or an eternal mystery. But surely, a priori, one should expect the world to be chaotic. One might… expect that the world evidenced itself as lawful only so far as we grasp it in an orderly fashion. On the other hand, the kind of order created, for example, by Newton's gravitational theory is of a very different character… Therein lays the "miracle" which becomes more and more evident as our knowledge develops… And here is the weak point of positivists and professional atheists, who feel happy, became they think that they have preempted not only the world of the divine but also of the miraculous…

As Stanley L. Jaki points out in his "<u>The ways to God and the roads of science</u>", Planck and Einstein, with their confidence in the reality of a causally connected and finite universe, provide extraordinary compelling evidence in favour of a realistic metaphysics and epistemology, midway between idealism and positivism.

LETTER TO PHYSICS TODAY

Dear Editor:

As usual, the "Search and Discovery" piece by B. Schwarzschild (*Physics Today Dec. 2011, pp 14-17*) on the 2011 Nobel prize was compact and informative. In Fig. **2b** a picture with confidence contours for $\Lambda > 0$ in the Ω_m, Ω_Λ plane shows convergence near $\Omega_m = 1/4$, $\Omega_\Lambda = 3/4$, using information inferred from high-z type supernovae, cosmic microwave background and baryon acoustic oscillations. It implies that there may be three times as much dark energy density as visible plus dark matter energy density.

I would like to point out that an alternative interpretation is possible using the Ω_m, Ω_k plane (where k <0 stands for space-time curvature) instead of the the

Ω_m,Ω_Λ plane. It is well known that Friedmann's solutions of Einstein cosmological equations for closed (k>0), flat (k=0) and open (k<0) universes were obtained assuming $\Lambda=0$ (probably because Friedmann realized that k<0 and $\Lambda>0$ play similar roles). Hence $\Omega_m+\Omega_k =1$ in the Ω_m,Ω_k plane results, in exactly the same downwards straight line from (0,1) to (1,0) as it does (in Fig. **2b**) in a flat (k=0) universe. It is easy to check that using the Friedmann-Lemaître solutions for an open universe with $\Lambda=0$, both Ω_m and Ω_k are <u>time-dependent</u>. Therefore the straight line from (0,1) to (1,0) describes in this case cosmic evolution from $\Omega_m \cong$ 1 (the big bang) to $\Omega_m \cong 0.044$ (now, t = 13.7 Gyrs) and beyond, with $\Omega_m=0$ in the distant future.

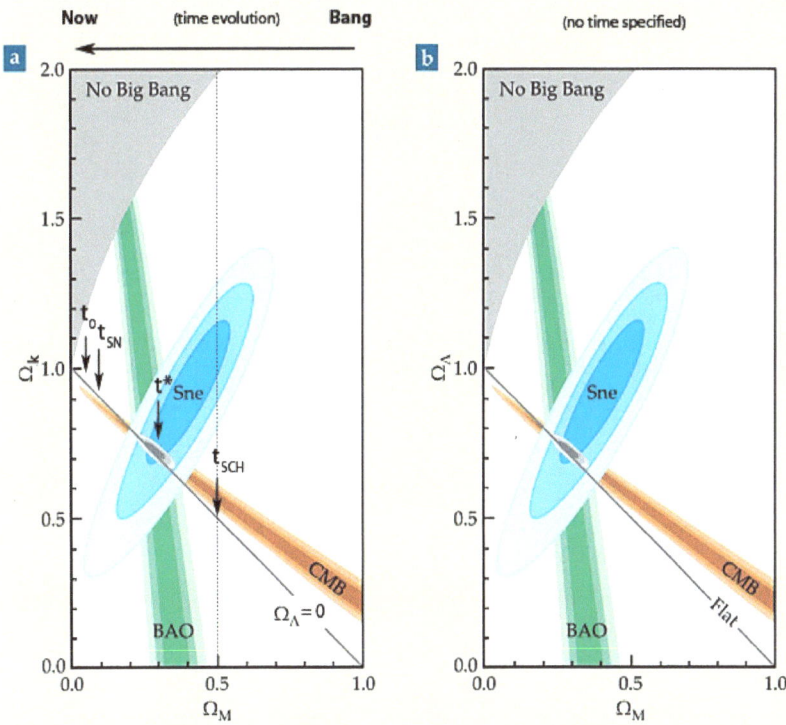

Figure 2: (a) Ω_k *vs.* Ω_m, and (b) Ω_Λ *vs.* Ω_m.

Light coming from high-z supernovae with $z \cong 1$ is arriving to us <u>now</u> but may have been emitted at a time somewhere before t = $t_{SN} \cong 7.5$ Gyrs ago, corresponding to $\Omega_m(t_{SN}) \cong 0.10$, and so forth for more distant supernovae.

Beyond some $z<z_+ = 20.7$ (cosmic Schwarzchild radius (time)) no galaxies (no stars) are observed for obvious reasons. To correlate quantitatively $\Omega_m(t)$ with $z(t)$ and with t itself is presently straightforward having into account that the relevant cosmic quantities (T_0=2.726 K $\pm$ 0.01%, t_0= 13.7 Gyrs $\pm$ 2%, H_0= 67 Km s^{-1}Mpc^{-1} $\pm$ 5%) are known with fair precision, after the COBE and WMAP satellites, as well as knowing of the present cosmic equation of state (RT= const.) and that radiation and matter densities become equal at atom formation ($t_{af} \cong 1$ Myr, $T_{af} \cong 3000$ K, $H_{af} \cong 6.48\times10^5$ Km s^{-1}Mpc^{-1}).

In other words this alternative interpretation might well be useful to substantiate the elusive "dark matter"-"dark energy" problem.

Yours,

Julio A. Gonzalo (Julio.gonzalo@uam.es)

UAM/U. San Pablo CEU, Madrid

CONFLICT OF INTEREST

The author(s) confirm that this chapter content has no conflicts of interest.

ACKNOWLEDGEMENT

Declared none.

REFERENCES

[1]   Gonzalo. J.A "Cosmic Paradoxes "(World Scientific: Singapore, 2012).
[2]   Gonzalo. J.A "Inflationary Cosmology Revisited" (World Scientific: Singapore, 2005).

CHAPTER 12

Why is the Universe the Way it is?

Manuel M. Carreira S.J.

Universidad Pontificia de Comillas, Madrid

Abstract: Every thinking person seeks answers to the question: Why is the Universe the way it is? Einstein asked himself: Did God have any freedom in choosing the initial parameters when creating? The "Anthropic Principle" presents Man as the conditioning factor to explain why the universe is the way it is. Change, design, contingency, creation and finality in connection with the question: Why the Universe is the way it is?

Keywords: Anthropic principle, Large Numbers Hypothesis, Chance, design, contingency, creation, finality.

There are questions, which, through human history, have engaged the interest of every thinking person, questions regarding our very being and existence and our relationship with the world in which we find ourselves They are now particularly pressing in the context of scientific developments of the last century which have a bearing on the possible origin of the Universe and its age, the origin of life on Earth and the possibility of finding life also in extra-terrestrial settings, the nature of intelligence and its development. These are no longer just subjects for abstract theological or philosophical discussions, but they are approached with data from the experimental sciences. Its remarkable that authors dealing with such topics are, in most cases, professional scientists who write both in research journals and also in those aimed at a learned but mostly humanistic public Books of this nature have recently become best sellers in almost every country, even in those which due to their social or economic hardships, might seem less likely to devote any interest to such impractical concerns.

For the past thirty-odd years I had the satisfying experience of finding audiences ranging from nuclear researchers to theologians, and from university professors to

---

*****Address correspondence to Manuel M. Carreira:** Universidad Pontificia de Comillas, Madrid; Tel 34-91-540-6101; Fax 34-91-372-0218; E-mail: ecarreira@res.upco.es

high school students who came to lectures meant to enlarge their viewpoints, not by pitting a way of knowledge against another, but by enriching their perspective with data and thoughts from other sources always seeking to better grasp a reality forever richer than all our efforts. In the US, in South America, in Europe, people are constantly surprised by the attendance and the interest shown whenever such subjects are presented. The complexity of human experience and the vastness of the mysterious material environment in which we find ourselves, from subatomic particles to galaxies, calls upon every thinking person to seek answers at deeper levels than what the common daily experience car offer, or our imagination, based on that experience, can pretend to represent.

It is proper of all knowledge to reduce the complexity of phenomena to patterns and relationships that find order in the vast collections of data from the different sciences and from personal experiences. There is no true development without this quest for understanding, ultimately seeking answers to the why, how and what for, of every level of existence This is the meaning of being rational and it is also traditionally expressed with the statement that we are all of us, at the deepest level, philosophers. Even the youngest child, as soon as the minimum command of language allows it, will begin the endless quest for reasons with the constant why that requires an answer deeper than the irrational "just because" with which adults will sometimes try to stifle the insistent search of a mind made for truth and logical clarity.

Nowhere is the question more pressing than in the realm of Cosmology What is the relationship between our existence and the existence and properties of the immense Universe where we are born? If Physics and Astronomy unequivocally point to a sudden beginning of the material world and to its future destruction down to the atomic level, what is the purpose of it all? Is it ultimately absurd, making our own life totally meaningless? Is all the greatness and beauty of so many wonderful people no more lasting than a passing ripple in a pond or the brief splendour of a wild flower?

To say that everything is the fruit of blind chance is to fall back upon the "just because" that fails to satisfy even the youngest child. Chance is not a scientific reason for anything, it is not a physical force and it is not a predictable pattern of development. It is, rather, the recognition of a lack of any true relationship between parameters or processes where our mind is striving to find some logical connection. It can never be a sufficient reason for orderly behaviour, even at the level of inanimate nature. And without order, there can be no science; at most, there will be a collection of facts, similar to the entries in a catalog or a social poll, sterile and incapable of development until some intuition of relationships gives structure to the bare data.

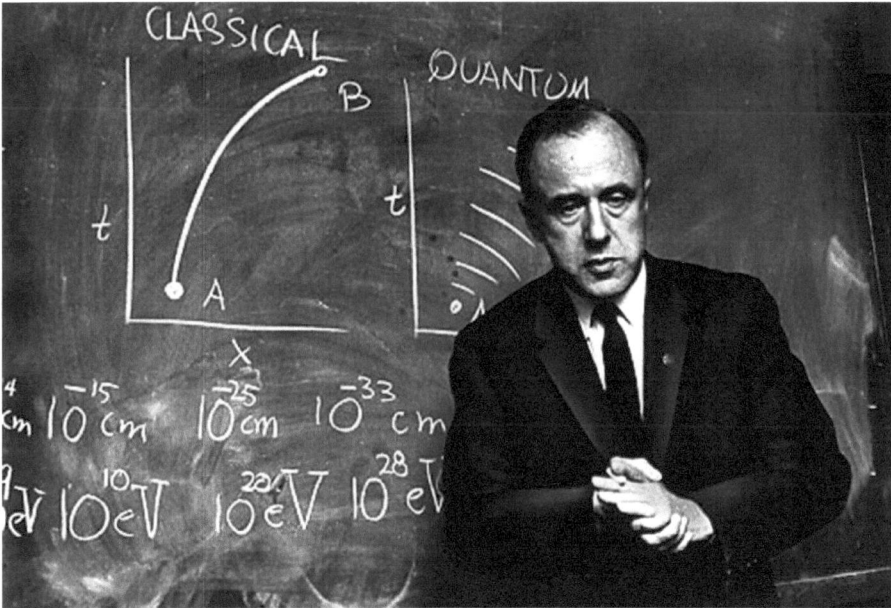

Figure 1: John A. Wheeler (1911-2008).

THE ANTHROPIC PRINCIPLE

The Copernican Principle, frequently invoked in modern Cosmology, stresses the homogeneity of the Universe: it denies to the Earth any special importance of position or properties that could be linked to the fact of human existence. Its name reflects Copernicus' proposal to place the Earth outside the geometrical center of the solar system, a suggestion previously made by Aristarchus and which was not accepted due to the problem of not observing stellar parallax (the change of position of stars during the year), rather than because of a philosophical evaluation of our existence on the planet. In the ancient approach to Cosmology, devoid of any idea of "forces" proper to matter, no reason could be given for the motions of heavenly bodies, whether one said that the Earth circled the Sun or just the opposite.

It was rather the abstract philosophical concept of "dignity" -and its consequence, "proper place" - that was invoked to justify the positions and motions of the different bodies, made of basic "elements" whose nature determined their rank. While the heavenly bodies were considered as formed of a "fifth essence", endowed with the highest perfection, all sub-lunar realities were described as composed of four different elements, fire, air, water and earth (a basic "dirt"). The properties of the fifth element included the perfect spherical shape, motion in circles at constant speed, and being unchangeable (incorruptible). Its proper place was the highest level in the skies.

The other material bodies were, in order of diminishing perfection, set at lower and lower levels: Fire at the highest, followed by air and water. At the bottom was the ball of dirt that constitutes the Earth, so that the heavier and more crass a body was, the more it seemed to seek the lowest point, from all directions. Left to itself in a deep well it should go to the center of our planet. The Earth was not in a place of honor, but rather at the bottom of the hierarchy of those "elements" proposed by ancient philosophical -rather than physical- ideas The center was only the place where a compass should be placed to draw orbits.

Copernicus proposed that the motions of the planets were simpler to describe by suggesting sun- centred orbits than with the Ptolemaic model. Still caught into the mind-set that all heavenly motions should be perfectly circular, he had to accept epicycles just as Ptolemy had, and his model was not accepted with any kind of enthusiasm at that time. Only the work of Kepler provided a really new viewpoint, obtained at the cost of a wrenching intellectual change: he had to abandon circles in favour of elliptical orbits, with changing speed and a Sun displaced from the center.

The next step was taken by Shapley over a century ago, when he showed that the Sun itself did not occupy the central spot in the Milky Way. It then became necessary to accept that other "Milky Ways" -galaxies- existed in great numbers spread in all directions in an immensity of mostly empty space. Today, the "finite but unlimited" Universe of Einstein's General Theory of Relativity implies *the impossibility of finding* a *center within 3-dimensional space* and affirms the uniformity and equal standing of all points. In fact, it is meaningless to even ask where we are exactly located within an expanding Universe with over 100 billion galaxies even; the Milky Way itself, our majestic cosmic city, fades into insignificance in such vastness. Size or position, by themselves, are not reasons to consider any given spot as privileged.

Nevertheless, beginning in the 1930s, we are faced with a rather interesting reaction: arguments are advanced, even more detailed and cogent, to further the view that we, as human beings, exist at a time and in a place which are in many ways atypical and, in that sense, privileged. This leads to the question of whether our existence is tied in a special way to properties or circumstances rather uncommon in the Universe. This question becomes especially significant when we consider the foreseeable consequences, according to physical laws, of even minute changes in the initial conditions of the Universe. We can echo Einstein's query *Did God have any freedom in choosing the initial parameters when creating?* We need to find a reason why the Universe exists, but also why it exists in such a way and with such properties that it can allow for the development of intelligent life. This is the basic thrust behind the formulation of the *Anthropic Principle*, which presents Man (understood in the philosophical sense as a "rational animal", independently of bodily shape and possible habitat) as the conditioning factor to explain why the Universe is the way it is.

The first suggestions of some ties between properties of the physical Universe at the present time and intelligent life are found in the non-dimensional relationships indicated by Eddington: the ratio of intensities between the electromagnetic force and the gravitational interaction, between the age of the Universe and the time for light to cross the classical electron radius, between the radius of the observable Universe and the size of a subatomic particle, all lead to the approximate value of 10^{40}. The number of existing nucleons is thought to be roughly equal to that same value squared. Is this some kind of childish numerology, or does it involve some serious meaning? The "Large Numbers hypothesis" suggests that Man can only exist in a context of time and place where such coincidences are found, even if no reason is given for the link.

Since the age of the Universe is necessarily changing, one can hold that intelligent life would be impossible in a different epoch, or it can be suggested that physical parameters are a function of time, but keeping the same relationships at least for as long as intelligent life exists. A different approach looks for those values to be related in this way at a particular and critical moment in the existence and evolution of the Universe, instead of the present time. In any case, it seems arbitrary to choose one proposal over another, especially when we have no experimental evidence to back any theory that includes a changing value for the charge and mass of the electron or for the intensity of gravitation. Such a theory was advanced by Dirac as a feature of his non evolutionary cosmology, and the idea of variable physical parameters (like the speed of light in a vacuum) is also proposed today without a direct bearing upon anthropic considerations.

In 1961 Dicke called attention to the fact that the ratios previously described apply to a Universe in a particular stage of its evolution, when intelligent life becomes possible only after about 10^{10} years, can there be heavy elements in such abundance that a planet like Earth can exist and develop the complex chemistry which allows life to arise and reach the level of intelligence. But the cosmic age must not be so great that stars capable of providing energy for a life bearing planet will have ceased to exist. The world we observe today is conditioned by the need to allow for the existence of the observer: this is an almost tautological statement, which was later dubbed the "Weak Anthropic Principle". In other words: since we would not exist at all if the Universe were significantly different, our activity as observers presupposes a physical environment suited for Man. This is really a very peculiar argument, equivalent to saying that the reason the atmosphere has oxygen is that otherwise we wouldn't be here to ask the question.

Collins and Hawking pointed out in 1973 a new consequence regarding the initial conditions of the Big Bang: only if the Universe had a density practically identical to the critical value would galaxies, stars and planets form from the original cloud. Thus any universe will need to be isotropic in order to allow for intelligent observers to exist. Carter, in 1974, extended the same reasoning to other initial conditions: any change in the basic parameters that define matter (density, interaction constants) would lead to consequences incompatible with physical evolution up to the human level. Thus the Universe, from the very first instant, *must* be such that it will allow for evolution towards intelligent life and for its actual existence at some time in history. This Is the "Strong Anthropic Principle", which includes a hint of *finality* even if the purpose of anything cannot be determined *in* any experiment nor can it be introduced into an equation.

Subsequent developments, due to Gale, Carr, Rees and Wheeler, underline in detail the many "coincidences" that must be present for stars to have a stable energy output during times long enough for life to develop and evolve, also for Carbon to be synthesized without changing totally into Oxygen (detailed energy levels in the Carbon and Oxygen atoms, predicted by Hoyle in 1953 and confirmed by Fowler)); for supernovae to explode and seed space with elements heavier than He. In every case the initial values of the four forces of nature and the total mass of the Universe control critical processes.

At a more local level we should point to the many circumstances, apparently very improbable and not predictable as a consequence of any physical law, that make our Earth a very special planet: its orbital radius places it in the very center of the habitable zone around the Sun; its mass (equal to the masses of Mercury, Venus. Mars and the Moon combined) is such that a moderate atmosphere can be retained. The Earth-Moon system -the result of a most improbable early collision with a planet larger than Mars) explains the stability of the tilt of the Earth's axis and the same collision gave the Earth an extra mass of hot iron which is the reason for the magnetic field that protects us from cosmic radiation, drives plate tectonics, explaining mountain building, ocean basins and volcanic activity (that recycles the crust and the atmosphere).

Even the catastrophic instances of widespread extinction (that eliminated 90% of all living forms) seem to be fortuitous but critical for an evolution that leads to mammals and, ultimately, Man. Any change in the planets history could have meant its unsuitability for life or the drastic limitation of living species. Not surprisingly. These views lead to a rather pessimistic outlook regarding the probability of intelligent life outside the Earth, even in the unimaginable vastness of the Milky Way.

Summing up the diverse ways to state the Anthropic Principle, we may look for the common thread, the Universe has some properties -both at its origin and in its evolution- seemingly not imposed by any previous *physical* necessity, that make intelligent life possible, at least in our planet. If we now ask why this happens, we can get basically two possible answers, either the Universe is the way it is *by chance* or it has been *designed* for our existence We must study the implications of each, while remembering that we are dealing with *metaphysical* problems since neither chance nor design can be experimentally proved or rejected, *scientific methodology cannot provide an answer* The scientific method deals *only with measurable parameters of material activity,* and can say nothing at all about finality, ethics, human freedom, art or many other important aspects of life

This fact is frequently forgotten in current discussions, where it is implicitly or explicitly argued that chance is a truly scientific answer and that design is only a religious presumption that science must reject. But chance is not a measurable quantity detected by any instrument in a lab. We should do a critical appraisal of its true meaning

I - CHANCE

The very idea of chance is tied to the *probability of several alternatives* in the context of many similar happenings, it has no meaning when we deal with an only instance Thus it is inapplicable to the initial conditions of the Universe, which by definition is *the totality of physical existence* It is to be expected, therefore, that the idea of a chance determination of parameters to explain the suitability for intelligent life is offered in the context of theories that hold the reality of an infinite number of "universes" that exist successively or simultaneously, even if they are *unobservable.*

The overwhelming majority of them are expected to be sterile for life, since changes in properties that render them unsuitable are much more probable than the simultaneous coincidence of all necessary parameters But in a *strictly infinite set* all possibilities must be realized, including the Universe we inhabit: our existence is the logical consequence of the endless variability of initial conditions The properties of the Universe are not aimed at the existence of Man. but necessarily lead *to it at some time and in some place.*

A successive Infinity of universes is also a favoured trick to avoid dealing with the problems of a beginning and an end. If the Universe is spatially "dosed", there will be an end of Its expansion, followed by a collapse. The stage is then set for a rebound with a new Big Bang at the end of each cycle, where all physical properties, from the dimensionality of space to the number and types of forces and particles, can be changed again and again in every possible way.

It is unnecessary to dwell upon the details of this hypothesis which is totally gratuitous neither the experimental data nor any acceptable physical theory leads to the prediction of a collapse. The density of the Universe is less than its critical value, and the new data (from Supernovae detected at early times, in 1998) show rather that the expansion is accelerating, not slowing down If the collapse were to occur, all matter should be compressed into the ultimate black hole from which no physical law allows a rebound. There is no repetition of the standard Big Bang which is described as taking place without any previous low density stage or any pre-existing space in which to expand: in the case of a cyclic universe there is a surrounding space where the event horizon delimits the possibility of motion for particles and energy.

Nor is the collapse equally applicable to all forms of matter and energy the outward motion of galaxies can be slowed down by gravitational attraction thusto an eventual stop and inversion of their trajectories, but radiation cannot be slowed or stopped in each cycle there will be a lower percentage of mass in the form of particles and more in the form *of* radiant energy as the result of stellar evolution An infinity of previous cycles in the distant past is thus incompatible with the present state of the Universe and its actual entropy and would lead, in finite time toan open Universe. Tolman has developed these arguments already than sixty years ago. One cannot accept either proposal as an explanation for the real neither Universe nor Hawking's theory of *a* system with no beginning or end because it is cyclic in imaginary time. *The* introduction of this variable can be satisfactory as a means to obtain *formal solutions of the equation*s of Relativity, but Hawking himself admits that the Universe did have a beginning in real time, and no mathematical formalism can counter the experimental data of cosmic density and other parameters that dearly lead to the denial of a future collapse.

When the infinity of universes is considered to exist simultaneously, the starting point is found in speculative unification theories of the four known forces of nature especially in the inflationary hypothesis of Guth and Linde. From the viewpoint of quantum physics the physical vacuum is described as the seat of an endless activity of spontaneous formation and destruction of all kinds of particles, with masses and other properties in infinite variety. During the Big Bang quantum fluctuations cause "seeds of universes", with all possible parameters to break off the chaotic substrate, each one developing independently and then expanding and evolving without any mutual interaction It is considered possible, at least theoretically that a sufficiently advanced technology might "create" such universes in the laboratory with just a few kilograms of mass compressed to the density of the initial physical vacuum.

Since no possible experimental check can validate these ideas, they are left outside the limits of strict scientific methodology, nor is there any measurement or observable consequence to support them They might be consistent with a mathematical formalism used in unification theories, which are also without any experimental proof so far, but such infinity of universes does not deserve to be discussed as a theory of the real world. Instead of solving a problem, it leaves unanswered the most pressing questions: *why each of those hypothetical universes in fact exists, why the physical vacuum is endowed with such quantum properties, why there would be a real infinity of material objects, which are not physically justified in any way.* Let us always remember that the physical sciences require the possibility, at least in principle, of an experimental check by some type of observation and measurement that can only deal with interactions taking place in this Universe

II-DESIGN

The only alternative left, as an explicit answer to the question of why the Universe is suitable for intelligent life, is to admit that its parameters have been designed for that purpose. This means that the concept of finality enters into the discussion: an idea which expresses something neither directly observable nor quantifiable, and which cannot be explained in terms of any mathematical equation or the four forces of nature. We are clearly dealing with something outside the realm of

Physics, because finality cannot lead to any experimentally verifiable prediction; we are in the field of *Metaphysics*, even if the data that lead to this discussion arise from the study of matter at all levels But this is simply to recognize that Physics does not describe the totality of what is knowable, a fact that needs to be accepted as applying to every level of human knowledge.

It will be most interesting to present the reasoning developed in the article "The Universe as Home for Man" by John Archibald Wheeler, one of the most highly regarded physicists of our time, stating the following chain of logical steps that leads to his *Participatory Anthropic Principle:*

- The most universal and basic property of matter is its mutability: it can be in many different ways

- Mutability presupposes adjustability, the capacity to be modified in multiple aspects.

- But whatever can be adjusted MUST be adjusted to be in a particular way rather than another

- Therefore the Universe had to be adjusted from the very first moment of its existence in order for it to be the way it actually is.

To put the argument in a very simple practical way there has to be a reason why the Universe is the way it is, making it real in this unique way instead of any other of the infinite variety of ways that were a *priori* equally possible. And because the most demanding and restrictive adjustment is the one required to make the Universe suitable for intelligent life, all physical parameters must be finely tuned from the first moment with a view to make possible the existence of Man (understood as a *rational animal*, without specifying its detailed structures).

The next step is to ask who the author of this initial "adjustment" is. If we also require an answer to the previous question, "Why does the Universe exist at all?" we will have to say that either the Universe was created or that its existence is justified with a childish "just because". Wheeler does not answer that basic problem, even if he states that that is the most important question of all. and its only logical answer has to be that there was an infinite Power that bridged the gap between nothingness and existence Wheeler himself stresses that we must explain the step from NOTHING to the universe, not from a supposed "physical vacuum" full of energy and endowed with quantum properties. Such a Creator gives a reason that will be logically sufficient also for the choice of initial conditions and parameters, since the Omnipotent has to be intelligent to know how to make a Universe and must have a reason to select its properties.

But Wheeler appeals instead to the concept of the "quantum observer" who causes the collapse of the wave function that describes a physical system and makes "real" one of its possible states. He then arrives at the most surprising kind of circular causality Man, by knowing the Universe, determines its initial state in such a way that it will lead to Man's existence, so that Man might be responsible for the Universe itself! We are really faced with a type of reasoning that has never been accepted in any science, even if it draws its basic approach from quantum mechanical experiments of delayed choice, which is explained by some physicists in terms of causality towards the past. But even in this view, such causality is never presented as conditioning the very existence of the observer responsible for it.

Wheeler holds that a universe is "real" only when it is observed, establishing a causal link between the unit of information and the reality it describes ("it from bit"), and turning around the normal view that knowledge presupposes the object that is known. He does not define the word "real", nor does he explain who is responsible for the observing process or when does it take place. It seems quite arbitrary to say that "Man" is the observer when, even in our scientific age, the near totality of mankind would be unable to understand what it must observe regarding the Universe, or what physical properties need to be adjusted in the distant past of 14 billion years ago. It would not be logical either to deny "reality" to the evolutionary stages preceding Man, stages that make up the almost totality of cosmic time. We end up with just some play of words or with an idealism that holds that my knowing is the determining factor to make things exist; in this case, the obvious consequence would be to say that NOW the material world exists because of my conscious activity, but not to conclude that it existed in a remote past before the observer performed its role.

Again, we should insist upon the fact that no answer is given to the truly basic question which Wheeler himself admits is at the core of the problem: *Why does something exist instead of nothing?* No quantum observer can be responsible for the fact that there is matter that obeys quantum laws, as well as the rest of all physical laws. Similarly, Stephen Hawking wonders why, *in fact,* there is a Universe that fits the equations that describe it as simply *possible* We are now at the point where we need to tackle the essential concept: what Wheeler expresses in terms of "mutability" is the metaphysical idea of CONTINGENCY, the essential inability of all changeable things to be a sufficient reason for their own existence Only a necessary Being, immutable, immaterial, unlimited in every respect, can exist by itself and be the sufficient reason for the existence of a non-necessary or contingent reality.

We are thus led to the last possible interpretation of the Anthropic Principle: the Universe was adjusted by its Creator, from the very first instant, with the purpose of making it such that its evolution would lead to the development of conditions suitable for life and its highest level, found in consciousness and intelligence. There is, then a sufficient reason why "something exists instead of nothing": the Creator is, ultimately, contemplating the existence of intelligent beings within the realm of the material world. It is truly surprising that the physical sciences lead us to this conclusion

CREATION AND FINALITY

Whoever acts intelligently, acts for some purpose, known and sought after; this purpose will determine the means to be employed to obtain it The Creator, infinitely powerful and capable of giving existence to the Universe, in a drastic step from nothingness to something, must know all the potentialities of the infinite possible worlds in order to choose from them the one that is suited for a given purpose.

We are dealing with a *free choice,* because creation in the strictest sense is not to be confused with some hypothetical "emanation" imposed by some kind of internal development, nor with a "dialectical" process of a creator who is not, ultimately, truly different from its creation. The true Creator, necessary and essentially unchanging and unchangeable, gives existence to a reality of a different and lesser order. This idea, which is found only in biblical theology among all the ancient cultures, is the unique way to avoid an infinite series of causes preceding the scientific Big Bang.

An infinite intelligence is required to know all the possible consequences of every minute change in physical parameters, during the history of each individual particle and combinations of particles. And a free will must select one of the possible sets of physical properties and laws in order to attain the chosen end, foreseen with certainty as the result of the activity with which matter is endowed at the very moment of its creation. The Creator will not need to intervene to make up any failings in the development of its original work, nor can any event be an unexpected surprise for the one who keeps everything in existence, moment after moment. Otherwise, that which cannot exist by itself at the beginning, would collapse to nothing at the moment it is left to its own contingency.

Pagels remarks that the Anthropic Principle is the closest that some atheist scientists dare to come to the concept of a creating God, but that in its logical incompleteness the principle fails both as a scientific and as a philosophical answer. He then adds that one can be more explicit and to the point with the *Theistic Anthropic Principle:* the Universe seems to be tailor-made for Man because it has, in fact, been made FOR Man. We do not get a proof from science regarding the existence of God the Creator, but we do get the data from which a *metaphysical* reasoning process leads to Him.

And the Creator we find is not an abstract concept of cosmic totality or personified "Nature" in a mythological sense, nor a God who creates as a banal exercise of power without regard for the beings thus created. Instead, we find a personal God, intelligent and free, whose creating action is ultimately an act of good will and love, not forcing in any way the creative activity, but giving it a sufficient reason, because goodness tends to be shared.

Only from this viewpoint we can also justify the existence of a Universe whose future evolution *necessarily leads to the destruction of all conditions and structures needed for life to exist.* To keep the Universe from being "a bad joke" it is necessary to find a way to save from meaningless futility the existence of Man that was presented as its justification. In a personal relationship of Man with God the whole of Creation turns to its Creator, because Man, intelligent and free, is an "Image and Likeness" of the Creator, capable of knowing and being thankful for the fact of existence and for all the levels of reality that make it possible. In this acknowledgment we find something new and qualitatively superior to the simple beauty of fireworks in stars and galaxies, which have fulfilled their purpose by preparing the environment for the arrival of Man.

Even so, human existence will appear as ultimately worthless if it is only something passing and destined to a final destruction. The complete answer must take into account the presence in Man of a new type of activity, *not reducible to the four forces that define matter.* This activity, in the form of *consciousness, abstract thought, free will,* can only be adequately explained by admitting a non-material reality as its source, even if this reality is intimately joined to a material structure and depends upon it for the way it acts. But whatever is non-material may, at least in principle, continue to exist even if matter decays. We thus avoid the bitter conclusion that a materialistic view imposes, as stated by Steven Weinberg at the end of his beautiful book "The First Three Minutes": *The more we know the Universe, the more absurd it looks.*

Neither Physics nor Metaphysics can lead us any farther. But a way appears open to save the cosmos from the ultimate absurdity of lacking a purpose: the existence of the material Universe has made possible the existence of this new non-material reality, which will not be constrained by the framework of space and time typical of matter, and thus it may survive the destruction of material structures in the unimaginable eons of cosmic evolution.

A complete answer will lead us even further. Human nature is not adequately represented or saved from destruction by considering simply its non-material component, incorruptible by physical processes and thus able to continue existing outside of the space-time framework where matter develops its activity Our body Is also an essential part of our being, and its future disintegration seems Incompatible with a true human survival. But no law of nature can account for any process that will preserve indefinitely the conditions necessary for organic life, nor is it possible for science to explain the survival of each Individual after death and corruption.

Philosophy itself seems powerless when confronted with this problem; it is only at the level of Theology, based upon knowledge given to us In revelation, that we find a satisfactory answer: the plan of the Creator includes a *rebirth* of each human being after death. A new way of existing, even for the body itself, will place human nature *outside of the limitations of space and time.* Thus matter is also saved from destruction, and, in the Christian view of Creation, it is so wonderfully honoured that in the person of Christ It has reached the very center of the Godhead. In the words of St. John and St. Paul: *"All things were created for Him: In Him everything was created, in heaven and on Earth, things visible and invisible"* (Jn 1,3.10, Col 1,16) and thus all of creation is saved from futility.

Anthropic considerations - St Peter Chrysologus (380 - 450) - Sermo 148: PL. 52, 596:

> *"Why do you ask how you were created and do not seek to know why you were made? Was not this entire visible Universe made for your dwelling? It was for you that the light dispelled the overshadowing gloom; for your sake was the night regulated and the day measured, and for you were the heavens embellished with the varying brilliance of the Sun. the Moon and the stars The Earth was adorned with flowers, groves and fruit; and the constant marvellous variety of lovely living things was created in the air, the fields, and the sea for you".* *"He has made you in his image that you might in your person make the invisible Creator present on Earth. He has made you his legate, so that the vast empire of the world might have the Lord's representative. Then in his mercy God assumed what He made in you; He wanted now to be truly manifest in Man, just as He had wished to be revealed in Man as in an image. Now He would be in reality what He had submitted to be in symbol".*

CONFLICT OF INTEREST

The author(s) confirm that this chapter content has no conflicts of interest.

ACKNOWLEDGEMENT

Declared none.

BIBLIOGRAPHY

[1]   Barrow, J. and Tipler, F., "The Anthropic Cosmological Principle", Clarendon Press, Oxford 1986

[2]   Einstein, A., Selection of quotes in a commemorative article by Kenneth Brecher on the centennial of Einstein's birth; Nature, March 15, 1979

[3]   Carr, B.J. and Rees, M.J., "The Anthropic Principle and the Structure of the Physical World", Nature. April 12, 1979

[4]   Hawking, Stephen. "The Edge of Spacetime", New Scientist. August 16, 1984.

[5]   Rees, M J," The Anthropic Universe", New Scientist, August 6, 1987. TRIMBLE, V,"Cosmology Man's place in the Universe", American Scientist, Jan-Feb 1977.

[6]   Steven.W, "The First Three Minutes". Basic Books, New York, 1977.

[7]   Wheeler. J A., "The Universe as Home for Man", American Scientist, Nov-Dec 1974.

[8]   Russell, Stoeger, Coyne, eds Physics, Philosophy and Theology, 1995

[9]   Russell, Murphy, Isham. eds Quantum Cosmology

[10]   Russell, Murphy, Isham. eds Quantum Cosmology and the Laws of Nature, 1996 Russell, Murphy, Peacocke, eds Chaos and Complexity, 1995

[11]   M. Heller, The New Physics and a New Theology, 1996 (translated by Coyne, Giovannini and Slerotowicz).

EPILOGUE

THE ROLE OF CHANCE AND DESIGN AS SCIENTIFIC ALTERNATIVES

Time and again it is said that the Big Bang has to be explained without introducing "arbitrary" initial conditions in order to obtain the present state of the universe through its evolution according to physical laws. This is asking the impossible: *any* initial conditions can and should be described as arbitrary when there is no previous state from which to derive them. To begin with a amount of mass or another, with some strength of each interaction rather than a different value, with quantum properties in a vacuum or in a soup of quarks, *all that* might seem something to be chosen on the basis of "elegance" or mathematical simplicity, but it does not impose on the Creator how the universe should be; if there is true *nothingness,* from noting nothing will develop in a logical way. This means that the creative act is totally free and it incorporates all and only those parameters that will eventually lead to the entire cosmic panorama intended by the Creator. We will see "chance events" that can also appear as arbitrary" to our limited understanding, but chance is meaningless for the intellect that knows the totality of existence. In the famous words of Einstein, "God does not play dice". There is no true opposition between Science and Theology, as long as it is recognized that each human *effort to know the totally of* our world is limited by its methodology and the very restricted experience we have of our environment. This requires that many independent views be considered as partial answers to the total question of where we are, how did we come to be, where is the universe headed and what the future means for us. Complementing each other, Science in all its richness, Philosophy and Theology, contribute also to the understanding of the masterpiece of nature, the Thinking Animal.

Emmanuel M. Carreira, S. J., Ph. D. - Universidad Comillas – Madrid

SCIENCE DOES NOT EXCLUDE A CONTINGENT UNIVERSE

The universe is <u>not</u> a <u>random</u> (chaotic) development; it is <u>not</u> a <u>necessary</u> (deterministic) result, either. The universe is <u>contingent</u>, *i.e.* <u>freely created</u>, as taught the great medieval Christian natural philosophers, Albertus Magnus, Roger Bacon, Thomas Aquinas, Buenaventura and Duns Scotto.

No wonder they were the ones who opened the way to Buridan and Oresme, and later, to Copernicus, Galileo and Newton. And then to modern science.

Contemporary scientific cosmology is rooted in medieval natural philosophy. To some it may it look chaotic, but not because of Planck, or Einstein, or Friedmann, or Lemaitre. It looks so those who have managed to give the false impression that modern science is seriously indebted to agnosticism or even to atheism.

Science and religion are two separate fields of knowledge, but it is possible to find a common ground: the universe is <u>intelligible</u>, *i.e.* it is well done and it is accessible to man´s intellect. If the universe were merely chaotic it would be very difficult to understand modern science.

Julio A. Gonzalo - U. San Pablo - CEU, Madrid

INDEX

www.ingramcontent.com/pod-product-compliance
Lightning Source LLC
Chambersburg PA
CBHW041717210326
41598CB00007B/680